Bridge Demolition Engineering

Other Titles of Interest

Assessment of Performance of Vital Long-Span Bridges in the United States, edited by Richard J. Kratky (ASCE/SEI 2004). This report describes the types and levels of assessments necessary for capturing the lifetime performance of long-span bridges, emphasizing the distinction between performance and condition assessments. (ISBN 978-0-7844-0710-3)

Bridge Decks: Design, Construction, Rehabilitation, Replacement, by Neal Bettigole and Rita Robinson (ASCE/ASCE Press 1997). Bettigole and Robison provide a comprehensive reference for the evaluation, testing, selection, and examination of relevant design criteria and alternatives for bridge decks. (ISBN 978-0-7844-0223-8)

Bridge Safety and Reliability, edited by Dan M. Frangopol (ASCE/SEI 1999). This report covers a wide range of applications of modern structural safety and reliability concepts and methods to highway bridge engineering practice. (ISBN 978-0-7844-0442-3)

Handbook of Concrete Bridge Management, by Fernando A. Branco and Jorge de Brito (ASCE/ASCE Press 2004). Branco and de Brito provide complete coverage of concrete bridges from planning, design, repair, maintenance, and demolition, with a focus on small- and medium-size bridges. (ISBN 978-0-7844-0560-4)

ASCE Manuals and Reports on Engineering Practice No. 157

Bridge Demolition Engineering

Best Practices

Bridge Demolition Subcommittee of the
Temporary Works Committee of the
Construction Institute of the
American Society of Civil Engineers
Bridge Demolition Subcommittee

Edited by
Josh Crain, P.E., S.E.
Lisa Briggs, S.E.
Sam Kevern, P.E., S.E.
Chris Tollefson, P.E.

Published by the American Society of Civil Engineering

Cataloging-in-Publication Data on file with the Library of Congress

LCCN 2023054024

Published by American Society of Civil Engineers
1801 Alexander Bell Drive
Reston, Virginia 20191-4382
www.asce.org/bookstore | ascelibrary.org

***Errata:* Errata, if any, can be found at https://doi.org/10.1061/9780784485361.**

ISBN 978-0-7844-1619-8 (print)
ISBN 978-0-7844-8536-1 (PDF)
ISBN 978-0-7844-8537-8 (ePub)

Manufactured in the United States of America.
27 26 26 25 24 1 2 3 4 5

Photo credit: Fairfax Bridge Center Spans, courtesy of MoDOT Photos.

MANUALS AND REPORTS ON ENGINEERING PRACTICE

(As developed by the ASCE Technical Procedures Committee, July 1930, and revised March 1935, February 1962, and April 1982)

A manual or report in this series consists of an orderly presentation of facts on a particular subject, supplemented by an analysis of limitations and applications of these facts. It contains information useful to the average engineer in his or her everyday work, rather than findings that may be useful only occasionally or rarely. It is not in any sense a "standard," however, nor is it so elementary or so conclusive as to provide a "rule of thumb" for nonengineers.

Furthermore, material in this series, in distinction from a paper (which expresses only one person's observations or opinions), is the work of a committee or group selected to assemble and express information on a specific topic. As often as practicable, the committee is under the direction of one or more of the Technical Divisions and Councils, and the product evolved has been subjected to review by the Executive Committee of the Division or Council. As a step in the process of this review, proposed manuscripts are often brought before the members of the Technical Divisions and Councils for comment, which may serve as the basis for improvement. When published, each manual shows the names of the committees by which it was compiled and indicates clearly the several processes through which it has passed in review, so that its merit may be definitely understood.

In February 1962 (and revised in April 1982), the Board of Direction voted to establish a series titled "Manuals and Reports on Engineering Practice" to include the manuals published and authorized to date, future Manuals of Professional Practice, and Reports on Engineering Practice. All such manual or report material of the Society would have been refereed in a manner approved by the Board Committee on Publications and would be bound, with applicable discussion, in books similar to past manuals. Numbering would be consecutive and would be a continuation of present manual numbers. In some cases of joint committee reports, bypassing of journal publications may be authorized.

A list of available Manuals of Practice can be found at https://ascelibrary.org/page/books/s-mop.

CONTENTS

BRIDGE DEMOLITION SUBCOMMITTEE

Andrew Ritter, P.E. (Siefert Associates, LLC)
Colin O'Hearn, P.E. (Kinetic Demolition)
Daniel Waugh, P.E., PMP (JR Vinagro Corporation)
Dave Byers, P.E., Ph.D. (Genesis Structures)
Jericho Tumanguil, P.Eng. (Engineering Constructs)
Jordan Aro, P.E. (GZA)
Martin T. Reed, P.E. (Steamboat Structures)
Matt Tebo (Kiewit Corporation)
Michael Garlich, P.E. (Collins Engineers)
Michael Hass, P.E. (Collins Engineers)
Reuben Zylstra, P.E., P.Eng. (Foothills Bridge Co.)
Thomas Rabinko, P.E. (Siefert Associates, LLC)
Troy Wright, P.E. (Silverado Contractors, Inc.)

BLUE-RIBBON PANEL REVIEWERS

Brian Witte, P.E., *Chair*
AASHTO-NSBA Task Group 10—Erection
Vice President, Construction Engineering, Parsons Corporation

Francesco Russo, Ph.D., P.E., M.ASCE, *Vice Chair*
AASHTO-NSBA Task Group 13—Analysis of Steel Bridges
Founder and Principal, Russo Structural Services

Ted A. Kniazewycz, P.E., F.ASCE
AASHTO Committee on Bridges and Structures
Technical Committee on Construction
Structures Division Director, Tennessee Department of Transportation

Additional Feedback Provided by

American Association of State Highway and Transportation Officials T4
National Demolition Association
National Steel Bridge Alliance

PREFACE

There is a growing need to replace our existing infrastructure as our country's bridge inventory ages. To meet this need, safe and controlled bridge demolition planning is of the utmost importance.

Most engineers and owners would acknowledge that additional considerations are required to analyze a structure as it is being constructed. During partial stages of erection, there are unique and potentially governing load cases that need to be considered beyond the final design of the permanent structure. A growing number of resources are available that have been developed as engineering guides for bridge construction.

During demolition, the structure will again experience temporary load cases while the structural system is being removed, creating changes to the structure capacity and potentially creating instabilities. Despite the similarities between bridge erection and demolition engineering, there is currently little to no formal guidance on engineering for safe and controlled demolition. This lack of consensus requires engineers to rely heavily on engineering judgment, which results in widely varying demolition design criteria and construction practices.

This document is a combined effort of specialty engineers and contractors who regularly work on bridge demolition projects. Together, this group has assembled the information included herein to document the current state of practice in the industry, with hopes of raising the bar on future work. The intent of this document is to work toward establishing consistent industry standards, by setting minimum expectations for what is included in a safe and effective demolition plan. This document is also intended to be a resource for engineers (both those performing and those reviewing the demolition analysis) to recognize that a bridge coming out of service need not be held to the same design standards as a permanent structure.

ACKNOWLEDGMENTS

The Temporary Works Committee gratefully acknowledges the contribution made by the Bridge Demolition Subcommittee and contributors in the preparation of this book.

DISCLAIMER

Although the Temporary Works Committee does its best to ensure that any advice, recommendations, or information it may give, either in this publication or elsewhere, is accurate, no liability or responsibility of any kind (including liability for negligence) is accepted by the committee.

Readers should note that the references are subject to revision occasionally and should ensure that they have the latest versions.

CHAPTER 1
BRIDGE DEMOLITION ENGINEERING

1.1 NEED FOR AN ENGINEERED DEMOLITION PLAN

According to the US Department of Transportation, there are more than 590,000 highway bridges in the United States of varying age and condition. This figure does not include the number of railroad bridges in service. Updating, repairing, and replacing the national bridge inventory involves a variety of bridge demolition activities ranging from deck repair to complete structure replacement.

Demolition activities often occur near other structures, infrastructure, and public spaces. Poorly planned demolition activities have caused property damage, resulting in additional cost to the contractor and the owner to repair or replace the damaged property. A properly engineered demolition plan considers adjacent structures to minimize risk and impacts and provides protection where warranted.

Unfortunately, demolition activities have occasionally resulted in worker injuries and/or fatalities, emphasizing the need for properly engineered demolition plans and procedures. Even in the absence of injury, demolition accidents bring about added project costs, construction delays, traffic inconvenience, and adverse publicity for the contractor and the owner. This demonstrates the need for engineered demolition plans and work sequences tailored to the specific project.

1.2 DEFINING ROLES AND RESPONSIBILITIES

Each demolition project has its own unique challenges, limits, and requirements. Demolition activities are normally a part of a larger project and may be undertaken by the general contractor or a specialty subcontractor. The owner's engineer or selected consultant is typically responsible for establishing the overall project limits and preparing the contract documents, specifications, and any special provisions that may be required. Contract documents typically include a summary of the extents of the demolition activities, constraints on demolition work, traffic control and access requirements, and global sequencing of work for the project. The owner, or their representative, may provide some level of field oversight during the demolition activities as well.

The specific demolition plans, sequences, and methods are developed by the contractor and their engineer and submitted to the owner, as required by the contract documents. Aside from structural analysis experience, ideally the demolition engineer should have field experience and a knowledge of demolition equipment and techniques. In many respects, demolition engineering is the reverse of erection engineering, with similar concerns for work sequence and global stability, combined with the added challenges that come with aging infrastructure. A predemolition safety meeting may be warranted, especially for complex or critical demolition operations. The contractor is typically responsible for coordinating the predemolition meeting, like a preconstruction or prelift meeting typically used for critical construction operations. The predemolition meeting should cover the operation being performed, key personnel involved, safety plan, communication plan, and any other topic that is critical to the success of the operation.

Project-specific and owner-specific requirements will always govern demolition operations. It is important to always review the complete contract documents, along with state and local requirements, for any specific regulations and limitations. Existing resources are also available for guidance related to bridge demolition engineering, some of which are described in the next chapter. Design codes such as those from the American Association of State Highway and Transportation Officials or the American Railway Engineering and Maintenance-of-Way Association should be followed for any structure that is supporting live traffic load.

1.3 CHOOSING THE RIGHT DEMOLITION METHOD

Bridge demolition is typically performed using controlled methods, following a bridge demolition plan that adheres to necessary code or specification requirements and considers unique loading conditions, stability concerns, and safety issues. For example, explosive demolition is one type of controlled removal method that can be used when site conditions are appropriate, safety concerns for adjacent infrastructure or the public are minimal, and environmental concerns have been mitigated. See Chapter 5 for more specific information on bridge removal with explosives.

Various demolition methods and equipment types may be selected by the contractor. Considerations in selection include project size, scope, and location (open rural area, congested urban area, over rivers, or railroads), bridge materials (type and condition), material salvage and reuse, available contractor equipment and preferences, environmental constraints, cost and schedule, and countless others. Several of these are discussed in more detail in the following paragraphs.

The age of the bridge structure plays a major role in determining the most effective bridge demolition method. Bridges that are being removed are often aging structures with years of wear and fatigue, resulting in reduced structural capacity. In addition, existing material properties can vary significantly from those used in modern construction. Structural degradation and corrosion contribute to section loss and stability concerns and may preclude certain types of demolition activities. All these factors need to be considered as the contractor determines the preferred equipment and removal methods.

Similarly, the structure type plays an important role in determining the safest and most effective bridge demolition methods. Bridge types with adequate redundancy and stability open up opportunities for the removal of structural elements, such as the deck and secondary load path members, prior to full superstructure removal (Figure 1-1). Other bridge types can become unstable when a single element is removed.

Figure 1-1. Excavators performing deck demolition on a composite steel plate girder bridge.

Another factor to be considered in bridge demolition planning is site access. The demolition plan for a bridge in a rural area with easy access (Figure 1-2) will look different from the plan for a bridge over a large body of water that requires barge access to facilitate removal. In urban areas (Figure 1-3), there may be limitations on crane and truck access, as well as restrictions on noise from percussive equipment. Environmentally sensitive areas may preclude access to certain areas or have additional stipulations on noise, debris, or other potential pollutants. When there are limited constraints and easy access to the bridge site, bridges can typically be knocked down using excavators, cranes, or other heavy equipment. In most instances, however, bridge demolition requires significant engineering to dismantle the structure in a way that is safe for the contractor, the environment, and the public.

Once established, the selected means and methods for demolition, along with any supporting engineering reports or calculations, are then conveyed

Figure 1-2. Demolition in a rural environment.
Source: Omega III Demolition, LLC.

Figure 1-3. Demolition in an urban environment over an existing rail and highway.

to the contractor's field personnel through a written demolition plan and associated drawings. Typically, this information is also submitted to the owner for review. An on-site predemolition safety meeting will help facilitate communication between the engineers who designed the demolition plan and the field personnel who are responsible for executing the plans. Execution of a properly prepared demolition plan should result in a safe and efficient demolition process from start to completion.

1.3.1 Partial versus Complete Demolition

There is an important distinction between partial demolition and complete demolition. Complete demolition means the entire bridge structure is to be permanently taken out of service and therefore can be analyzed assuming that the remaining service life is finite and known.

Partial demolition often involves deck removal and replacement. In this scenario, the concrete deck is the only part of the bridge being removed. The remaining superstructure and substructure are reused and must be analyzed and protected to ensure that the service life of the rehabilitated bridge is not impacted by partial demolition activities. Although many of the same tools and load development principles are similar for both complete and partial demolition, the design limits may vary, and additional resources should be consulted, especially for demolition phases where the bridge remains open to traffic while supporting equipment.

1.4 ENGINEERING ANALYSIS CONSIDERATIONS

Analysis methods used in performing demolition engineering must be appropriate for the type and configuration of the structure being analyzed. Demolition projects may require a detailed step-by-step analysis, as both loads and load paths are modified from their as-built configuration. Supports may be added, altered, or removed, which can change structural behavior. Bearings may go into uplift and require counterweight to maintain stability. Tension-only and compression-only members may experience load reversal and necessitate strengthening. Hinges may become unstable and need to be fixed in place. In continuous structures, demolition can create load redistribution and may impact components some distance away from the immediate work location. Demolition may warrant taking advantage of nonlinear material properties and verifying second-order effects on a partially deconstructed structure.

Complex bridge demolition analyses require an experienced engineer to evaluate the assumptions used for boundary conditions, design loading, and material properties. Structural engineering software idealizes boundary conditions, which may make upper- and lower-bound analyses necessary to envelope real-life behavior. Material properties may be unknown, requiring engineering judgment or destructive testing to determine appropriate design values. The analysis must take into consideration the actual sequencing of the demolition, changes to boundary conditions, and the equipment and associated loading that the contractor intends to use.

Ultimately, it is up to the contractor and their engineer to determine the most effective demolition method and to develop a plan for the work to be performed safely, within the constraints of the project contract documents. This publication aims to provide resources and information to facilitate this goal.

1.5 ENSURING PUBLIC SAFETY

Bridge demolition is an industry that requires highly technical operations to be executed near the traveling public or other public spaces. Consistent with the ASCE Code of Ethics, the first and foremost goal of any demolition plan should be to ensure the safety and welfare of the public, especially the people performing the work. Demolition plans should include all information regarding the maintenance of traffic requirements, lane or road closures, restrictions, and durations necessary to protect public safety for all demolition operations over or adjacent to live traffic and public spaces.

CHAPTER 2
MANUALS AND SPECIFICATIONS FOR DEMOLITION

2.1 INTRODUCTION

Most new bridge projects and bridge rehabilitation projects require partial or complete demolition of existing bridge structures. Bridge demolition has become increasingly complex with increased population and traffic density, environmental concerns, and economic and safety impacts of road closures. Partial or complete structure removal is often executed on the critical path and, like bridge construction, carries safety risks resulting from temporary partial support conditions. Existing demands may also increase as portions of the bridge are removed and the stiffness of the structure changes.

Despite the concerns and associated risks, no unified bridge demolition specifications or guidelines exist. Typically, requirements and regulations are provided on a project- or owner-specific basis, resulting in guidelines that can vary substantially on a project-by-project basis. In many cases, there are no specific stipulations for demolition, leaving decisions regarding engineering analysis and appropriate methods for removal to the contractor.

Government owners are increasingly adopting bridge demolition criteria, as documented in *Bridge Demolition Practices* (Garlich and Simpson 2019). Private owners, such as the railroads and National Park Service, also have some documented requirements for bridge demolition and removal. This section lists several available references and guidelines for bridge removal. This is not a comprehensive list, and all information should be verified. It is up to the engineer to confirm that all local and federal requirements are being met, and that the appropriate version of all codes and standards is being used. In the absence of local or project-specific requirements, however, the sources listed subsequently may offer guidance.

2.2 DESIGN CODES

No current design codes exist that specifically address the analysis of bridges for/during demolition operations. As such, it is a common industry practice to extrapolate from existing design codes using the guidance of sound engineering judgment for analysis. Demolition operations, like bridge construction, may subject a structure to a state of partial support (increased unbraced length, loss of composite action, limited alternate load paths, and so on) when compared to a structure in the final designed condition. These temporary stages during removal often result in reduced structural capacity and may control the design of removal operations. The existing condition of the bridge also needs to account for capacity reduction resulting from corrosion or section loss.

When applying codes that were written for the final bridge design to bridge demolition, there are a few key differences to keep in mind. First, the equipment used for engineered bridge demolition is often known with a high level of certainty. Second, demolition operations typically take place over a finite window of time. This differs significantly from the assumptions that are used in all new bridge design specifications. The design of new bridges is based on assumed live loads, along with high load factors to account for the uncertainty in live load. New bridges must be designed for a service life that typically exceeds 50 years. This requires design checks to ensure that the structure will maintain performance during its lifetime. Simply put, bridge design codes are intended to protect the design of new structures on the basis of uncertain loads and long service life, whereas bridge demolition does not have the same concerns. Using design logic intended for new design may not be appropriate and may be overly conservative. Refer to Section 2.2.1.5 for further discussion on the use of AASHTO codes for demolition engineering.

2.2.1 American Association of State Highway and Transportation Officials

Two AASHTO documents are available that are typically used for bridge demolition: AASHTO *LRFD Bridge Design Specifications* (AASHTO 2020) and AASHTO *Manual for Bridge Evaluation* (AASHTO 2018) AASHTO also has a construction specification (AASHTO 2017a) that provides limited requirements for demolition. In addition, the AASHTO *Guide Design Specifications for Bridge Temporary Works* (AASHTO 2017b) would be applicable if temporary works are needed for bridge removal.

2.2.1.1 AASHTO Load and Resistance Factor Design Bridge Design Specifications. The AASHTO *LRFD Bridge Design Specification* is intended for the permanent design of highway structures; however, this document provides imperative guidance for evaluating the strength and stability of bridge structures. In essence, the evaluation of structural members for strength and stability during demolition is similar to the design/evaluation of structures for permanent use. The main differences for demolition operations lie in the design load factors, load combinations, load distribution, and necessary design checks.

2.2.1.2 AASHTO Manual for Bridge Evaluation. The AASHTO *Manual for Bridge Evaluation* (MBE) provides guidance on evaluating existing structures. Although the MBE usually references the AASHTO *Bridge Design Specifications* for specific analysis, it provides guidance on the load rating of existing structures, as well as assumed historic material types and capacities, which can be used when bridge plans are not available or complete. The MBE is often used for deck removal projects, where the superstructure and substructure will remain in service, or for checking bridge members that will be temporarily supporting construction equipment during demolition but will ultimately be taken out of service.

The MBE provides guidance on the basis of both the AASHTO *LRFD (Load and Resistance Factor Design) Bridge Design Specifications* and the AASHTO *Standard Specifications for Highway Bridges* (AASHTO 2002). The AASHTO *Standard Specification* allows evaluation using either LFD (load factor design) or ASD (allowable stress design) limit states. Evaluation using service loads and ASD limits states can be preferable, as other areas of demolition analysis (falsework design, pick weight determination for hoisted removal, and anticipated deflections) are often performed using service level loads.

2.2.1.3 AASHTO Load and Resistance Factor Design Bridge Construction Specifications. AASHTO's *LRFD Bridge Construction Specifications* provides limited requirements for bridge demolition. Although the specification requires the submittal of demolition drawings for structures or portions of structures, when removal is over or adjacent to public traffic or railway property, or if specified in the contract documents, the specification provides no requirements for the contents of those drawings.

2.2.1.4 AASHTO Guide Design Specifications for Bridge Temporary Works. The AASHTO *Guide Design Specifications for Bridge Temporary Works* provides guidelines for the design and construction of falsework, formwork, and temporary retaining structures for use in bridge construction, rehabilitation, and demolition. Information is provided for load combinations and load factors, construction loads, wind loads, along with additional loading information and general design guidelines.

2.2.1.5 Application of AASHTO Design Codes for Demolition. No specific load factors or live load distribution factors are prescribed for demolition equipment, nor load combinations that are required by code. It is often permissible to analyze construction equipment using the Strength II load combination when following AASHTO LRFD BDS for analysis. Strength II is defined as "Load combination relating to the use of the bridge by Owner-specified special design vehicles, evaluation permit vehicles, or both without wind." Well-defined demolition equipment with known weights is similar to permit vehicles.

Similarly, AASHTO MBE distinguishes between operating level and inventory level limit states. Operating Level is defined as "the maximum load level to which a structure may be subjected" and allows for temporary loading greater than the standard vehicle design loading. Inventory level loading is typical for standard traffic and allows for "comparisons with the capacity for new structures and, therefore, results in a live load which can safely utilize the existing structure for an indefinite period of time." Thus, in certain scenarios and when approved by the owner, it is permissible to perform bridge demolition analysis at the operating level limit state.

It is also important that structures with staged demolition sequencing (bridge structures supporting traffic loads during or after demolition operations) should be treated differently from structures that will no longer be in service. The engineer should always consider the unique circumstances of the structure when determining the appropriate load factors, combinations, and design limit states.

2.2.2 American Institute of Steel Construction, Steel Construction Manual

Although the AISC *Steel Construction Manual* (AISC 2017) does not specifically address bridge structures, it is referenced in AASHTO and provides guidance where AASHTO is silent or not applicable. It is common for historic bridge structures to fall outside the scope of the most recent AASHTO LRFD specification.

2.2.3 American Concrete Institute, ACI 318

ACI 318 (ACI 2019) does not specifically address bridge structures but provides additional guidance for concrete structures.

2.2.4 American Railway Engineering and Maintenance-of-Way Association

The governing specification for bridge structures supporting railroads is American Railway Engineering and Maintenance-of-Way Association

(AREMA) *Manual for Railway Engineering* (AREMA 2023), which should be considered when bridge demolition operations concern railroad bridges. "Construction load combinations, stresses, and stability" are covered in Part 26.4.4. This section is intended for bridge construction, but similar principles can be applied to bridge demolition. Temporary structures are covered in Part 28.

2.2.5 Bridge Construction Guides

2.2.5.1 AASHTO *Guide Specifications for Wind Loads on Bridges during Construction.* This guide specification is intended for the construction of bridges before the deck is in place. The same approach to wind loads can be used for bridge demolition projects where the deck is being removed. This document also provides guidance on reducing the maximum wind speed for design on the basis of the anticipated exposure window.

2.2.5.2 Federal Highway Administration *Engineering for Structural Stability in Bridge Construction.* This document (FHWA 2015) is targeted at evaluating bridges for stability during construction but also addresses demolition activities and provides some guidance for demolition plans and supporting calculations.

2.2.5.3 Wisconsin Department of Transportation *Development of a Bridge Construction Live Load Analysis Guide.* This guide (Garlich and Miller 2011) was developed for the analysis of construction loads on permanent bridge structures, including new construction and rehabilitation projects. The principles applied are similar to those of bridge demolition projects.

2.2.6 State-Specific Specifications

At least 26 states have some requirements related to bridge demolition that are mentioned in the individual states' standard specifications. The states included here provide more detailed requirements, plan requirements, and so on. It is always good practice to search for the individual state's requirements where the work is being performed prior to analysis. The Federal Highway Administration provides links to all state bridge specifications; links are updated frequently but are subject to change (FHWA 2023). It is the engineer's responsibility to ensure adherence to the most current specifications for each project on the basis of the project location, owner requirements, and methods used.

References to several owner-specific specifications can be found in Section 2.3, "Appendix." This is not a comprehensive list, and the engineer is responsible for validating information used for their specific project.

2.2.7 Railroad Agency Specifications

In addition to states and territories, cities and railroads may have additional requirements. Railroad requirements are often more stringent than those for Department of Transportation owners.

BNSF Railroad
Guidelines for Preparation of Bridge Preparation and Removal Plan over the BNSF Railway: https://www.bnsf.com/bnsf-resources/pdf/in-the-community/bnsf-demolition-guideline.pdf

CSX Railroad
Public Project Information: For Construction and Improvement of Projects That May Involve the Railroad: https://www.csx.com/index.cfm/library/files/about-us/property/public-project-manual/

Kansas City Southern Railroad
Guidelines for the Design and Construction of Railroad Overpasses and Underpasses: See Section III. https://ftp.dot.state.tx.us/pub/txdot-info/brg/kcsr_guidelines.pdf

Union Pacific Railroad
Guidelines for Preparation of a Bridge Demolition and Removal Plan for Structures Over Railroad: https://www.up.com/cs/groups/public/@uprr/@realestate/documents/up_pdf_nativedocs/pdf_up_reus_industry_demo.pdf

2.2.8 Current and Existing Research

In addition to the aforementioned references, several existing and current research projects are related to bridge construction and demolition or to the load types and conditions applied to bridge structures. Research papers can provide guidance on analysis approaches in areas where the current codes may be silent. The engineer should exercise caution and ensure adequate understanding of the research and the impacts it has on design when using research papers as a basis for analysis. Although research may provide guidance or recommendations, the engineer is still fully responsible for all engineering design decisions.

2.3 APPENDIX: OWNER-REFERENCED SPECIFICATIONS AND STANDARDS

The following owner-specific references are intended as a guide and are not a comprehensive list. This information is subject to change and the demolition engineer is responsible for verifying the current owner requirements.

Caltrans (California)

Bridge Construction Records and Procedures Manual, Volume II

For guidance specific to removal or demolition of existing structures, refer to the bridge construction memos 124-2, "Demolition Plan Review," and 124-3, "Demolition Over and/or Adjacent to Union Pacific Railroad Company Tracks,"

https://dot.ca.gov/programs/engineering-services/manuals/bridge-constr-records-proc-manual-vol2.

Construction Manual

See "Section 60 Existing Structures," 4-6003B Removing Bridges

https://dot.ca.gov/programs/construction/construction-manual/section-4-60-existing-structures.

Kansas

Kansas Department of Transportation Special Provisions to the Standard Specifications, 2007 Edition

The controlled demolition specification categorizes demolition projects by risk and complexity with different submittal criteria for each category.

https://www.ksdot.org/Assets/wwwksdotorg/bureaus/burConsMain/specprov/2007/pdf/07-07014.pdf.

New York State Department of Transportation

Standard Specifications (US Customary Units), Volume 2 of 4, Sections 200–599

The Structures section begins with Section 550. Also, see "Removal of Structural Concrete," Section 580 and "Removal of Structural Steel," Section 589.

https://www.dot.ny.gov/main/business-center/engineering/specifications/english-spec-repository/2020_5_specs_usc_tc_vol2.pdf.

Washington State

Standard Specifications for Road, Bridge, and Municipal Construction, M 41-10

Section 2-02 covers bridge and obstruction removal.

See Section 2-02.3(2)A "Bridge and Structure Removal" and Section 2-02.3(2)A3 "Use of Explosives for Bridge Demolition."

https://wsdot.wa.gov/publications/manuals/fulltext/M41-10/SS2024.pdf.

Wisconsin

Wisconsin Department of Transportation Roadway Standard Specifications

Section 203 covers the removal of old culverts and bridges.
https://wisconsindot.gov/rdwy/stndspec/ss-02-03.pdf.

US National Parks Service

National Parks Service has historic buildings and bridges within its purview. In addition to the demolition information, it has good information on historic building surveys, preservation, and other items related to historic structures.
https://www.nps.gov/dscw/del_constructiondb.htm.

Alberta, Canada

Standard Specifications for Bridge Construction
See Section 21, "Demolition, Disposal, and Salvage of Bridge Structures."
https://open.alberta.ca/dataset/1dce7c2c-5cba-4eca-8206-48be7ba6b95b/resource/e506b159-0eb2-4f1a-871a-cf2ac2ee1c46/download/standardspecificationsforbridgeconstruction2017.pdf.

Manitoba, Canada

Specifications for Bridge Construction: *Specifications for Demolition and Removal of Existing Structures*
Manitoba Infrastructure and Transportation's Specification 1020(i) covers the demolition and removal of existing structures.
https://www.gov.mb.ca/mti/contracts/pdf/manual/1020i.pdf

REFERENCES

AASHTO (American Association of State Highway and Transportation Officials). 2002. *Standard specifications for highway bridges*. 17th Ed. Washington, DC: AASHTO.

AASHTO. 2017a. *Construction handbook for bridge temporary works*. 2nd ed. Washington, DC: AASHTO.

AASHTO. 2017b. *Guide Design specifications for bridge temporary works*. 2nd ed. Washington, DC: AASHTO.

AASHTO. 2020. *LRFD bridge design specifications*. 9th ed. Washington, DC: AASHTO.

AASHTO. 2018. *Manual for bridge evaluation*. 3rd ed. Washington, DC: AASHTO.

ACI (American Concrete Institute). 2019. Building code requirements for structural concrete, ACI 318-19. Farmington Hills, MI: ACI.

AISC (Americn Institute of Steel Construction). 2023. Steel construction manual. 16th ed. Chicago: AISC.

AREMA (American Railway Engineering and Maintenance-of-Way Association). 2023. *Manual for railway engineering*. Lanham, MD: AREMA.

FHWA (Federal Highway Administration). 2015. *Engineering for structural stability in bridge construction*. FHWA-NHI-15-044. Washington, DC: FHWA.

FHWA. 2023. "State specifications." Accessed December 2, 2023. https://highways.dot.gov/federal-lands/specs/state-specifications.

Garlich, M., and S. Miller. 2011. *Development of a bridge construction live load analysis guide*. WisDOT ID No. 0092-10-13. Madison, WI: Wisconsin Dept. of Transportation.

Garlich, M., and J. Simpson. 2019. *Bridge demolition practices*. NCHRP Synthesis 536. Washington, DC: Transportation Research Board, National Research Council.

CHAPTER 3
EQUIPMENT AND TOOLS FOR BRIDGE DEMOLITION

3.1 INTRODUCTION

The tools or equipment required for a demolition project vary widely depending on structure type, site conditions, and project schedule. Often, the capabilities of available equipment and the type of structure being removed determine the means and methods of removal. Standard equipment is commercially available; however, specialty equipment, often fabricated by the contractor for a specific task, is sometimes required.

This chapter includes an overview of standard equipment used for bridge demolition projects. Standard equipment typically performs a wide range of operations common to most demolition jobs. These tasks can include structure removal (cutting, shearing, rubblizing), material processing, and hauling.

A section on specialty heavy lift equipment has also been included. This type of equipment may be required on complex demolition projects owing to access restrictions, schedule requirements, or sensitive environmental areas. Specialty equipment is often more expensive than standard equipment and may require specialty subcontractors. However, it has the benefit of performing removal operations with greater precisions and/or speed than standard equipment. The use of specialty equipment should be evaluated by the contractor and the engineer on a project-specific basis.

3.2 REMOVAL EQUIPMENT

3.2.1 Excavators

Excavators are used on nearly all demolition projects. Excavators typically perform operations such as earth work, structure removal, material movement, and processing of material. Most excavators are confirgured with similar components. Refer to Figure 3-1 for the parts that make up a common excavator.

Excavators come in various weight classes and may be configured with full or zero-swing counterweight (Figure 3-2). In addition, excavators come with a variety of boom lengths, stick lengths, components, and

Figure 3-1. Parts of an excavator.

Figure 3-2. Zero-tail swing excavator.

attachments. A summary of the varying excavator sizes/classes along with the typical work performed is described in the following.

Mini (0 to 10 metric tons or < 22,000 lb): These excavators are typically used when space is limited or if the load capacity of a structure supporting an excavator is of concern. Mini excavators are equipped with smaller attachments that work well for demolishing smaller or finer sections without damaging the remaining bridge structure elements. This size machine can be supported by most bridge structures open to public traffic.

Small (11 to 19 metric tons or 23,000 to 47,000 lb): Small excavators provide a substantial increase in power from the mini range but are still compact enough to offer many advantages like mini excavators. This weight class contains the most zero or near-zero/minimum tail swing radius models. The zero-tail swing excavator is convenient for bridge demolition work where space is at a premium. This size of the machine can typically be supported by most bridge structures open to public traffic.

Medium (20 to 35 metric tons or 48,000 to 77,000 lb): Medium-sized excavators provide the capacity and power to perform most bridge demolition projects. This class contains multiple zero-tail swing models. They can be equipped with the same attachments as the large excavators but are less powerful and may have slower production rates. This size machine will commonly require engineering to determine whether the structure can support the equipment weight.

Large (36 to 200 metric tons or 78,000 to 450,000 lb): Large excavators are the most powerful class of excavators that are used when the largest and most powerful demolition attachments are needed. An existing bridge structure can rarely support an excavator weighing more than 100,000 lb (45.3 t) without supplemental support or additional engineered systems. This weight class also includes specialty high-reach and long-reach excavators. High-reach excavators can achieve a vertical reach from 60 to 170 ft (18.3 to 51.8 m). Long-reach excavators can reach up to 100 ft (30.5 m) horizontally (Figure 3-3).

3.2.1.1 Second and Third Member Attachments

Second Member Attachments: These are tool attachments that are mounted directly on the boom of an excavator. This configuration allows for the use of heavier tools that can deliver more power. These tools may be required when cutting and lifting heavier/larger loads (Figure 3-4).

Third Member Attachments: These are tool attachments that are mounted on the stick of the excavator. Excavators with third member

Figure 3-3. "Large" high-reach excavator.
Source: Priestly Demolition.

Figure 3-4. Excavator with second member attachment connected directly to an excavator boom.
Source: Priestly Demolition.

Figure 3-5. Excavator with third member attachment connected to an excavator arm/stick.
Source: Priestly Demolition.

attachments are the standard configuration of the excavator. This configuration provides more reach and range of motion (Figure 3-5).

3.2.1.2 Excavator Tool Attachments Used for Demolition. Excavator tool attachments are chosen for the type of operation that will be performed. Although most tool attachments are commercially available, custom tool attachments may be used. An overview of common tool attachments is provided subsequently.

Bucket: Tool attachment used for digging, sorting, moving, and loading of material for transport.

Bucket and Thumb: Tool attachment that contains the standard bucket attachment and an additional hydraulic thumb, which is used to secure material in the bucket (Figure 3-6).

Slab Bucket: Specialized tool attachment specifically used to remove sections of deck or other bridge elements (Figure 3-7).

Grapple: Hydraulically articulated jaws that are typically used for managing debris or used to hold elements while they are being demolished. Sometimes this is referred to as a claw attachment. This attachment is also available as a rotation grapple (sorting grapple), which can rotate for additional flexibility (Figure 3-8).

Hydraulic Hammer or Hoe-Ram: Tool attachment that contains a hydraulically powered piston that is commonly used to break or rubblize concrete (Figure 3-9).

Figure 3-6. Excavator with a bucket and a thumb removing a barrier.

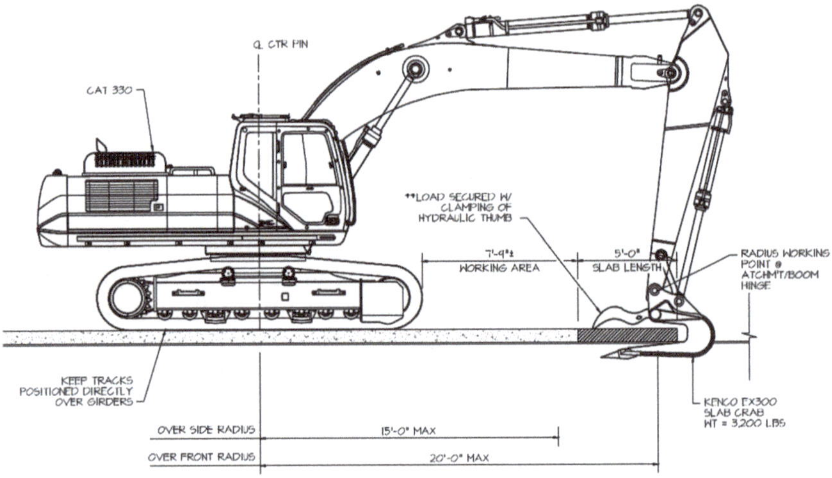

Figure 3-7. Slab bucket.

Processor: Hydraulically controlled jaws used to crush concrete and cut the reinforcing steel (Figure 3-10).

Shear: Hydraulically controlled jaws used to cut rebar and structural steel members. The thickness of steel that shears can cut is dependent on the specific shear attachment and size of excavator. Shears can also be used to pick existing members (Figure 3-11).

Pulverizer/Muncher: Tool attachment that is similar to a processor, used to process or crunch the concrete into smaller sections. These are mechanical and have no hydraulics. These are powered by only the bucket cylinder (Figure 3-12).

Figure 3-8. Rotating grapple attachment.

Figure 3-9. Excavator removing a deck with a hammer/breaker attachment.

3.2.2 Cranes

Various types and models of cranes are available with hoisting capacities that can range from 20 t up to 1,200 t. Cranes are used to hoist material, equipment, tools, and personnel.

Types of cranes typically used for bridge demolition include

- Lattice boom crawler,
- Hydraulic boom crawler,
- Rough terrain (RT) hydraulic,

Figure 3-10. Processor attachment (third member). Source: Priestly Demolition.

Figure 3-11. Shear attachment (second member) separating a deck and lifting a steel girder.

Figure 3-12. Pulverizer/muncher attachment.

- All-terrain (AT) hydraulic, and
- Lattice boom truck.

Cranes are chosen on the basis of each bridge demolition project's unique constraints and characteristics. The size and weight of the hoisted bridge elements are typically determined by the means and methods of the demolition operation. The weight of components that will be hoisted should be determined prior to the operation and should account for any uncertainties and the potential for any incidental loading. See Chapter 4, "Loads," for additional notes of the development of hoisted loads.

Lift Plan: When cranes are used for bridge demolition, a lift plan should be prepared as an integral part of the overall bridge demolition plan. The lift plan is an analytical tool used to document the proposed lift and ensure that it can be safely executed within the project constraints. Lift planning should be performed according to ASME P30.1 (ASME 2019) requirements and project-specific requirements. Lift planning takes into consideration adjacent structures, facilities, underground and overhead utilities for clearances, owner-specific factors of safety requirements, and loads induced by the crane support foundations.

It is common for owners to require lift plans as part of the demolition plans that must be submitted. Owners may also define a "critical lift" that has additional requirements, such as a critical lift plan and a critical lift planning meeting. Critical lifts are defined by the owner, typically when the pick weight exceeds a certain percent of the crane's safe working load, when two cranes are used to make a single pick, or when the crane is required to walk with the pick. It is the engineer's and contractor's responsibility to understand what the definition of a critical lift is and all the planning requirements and submittals for a critical lift.

Project Site: The project site and terrain should be evaluated to determine limitations on the choice of crane. Rough terrain cranes have limited capacity but are designed for traveling off-road with little site preparation (Figure 3-13). All-terrain cranes have a larger range in capacity but have limitations for operating off-road. Crawler cranes can travel on a variety of surfaces but are often heavier and require additional consideration for the surfaces that the crane will load.

In all cases, consideration should be paid to the requirement for the protection of existing roadways, sidewalks, or other infrastructure to remain. Cranes, in general, require a level, stable surface to operate from when hoisting material. Additional planning and/or preparation may be required to account for substandard soils, protection of pavement, or protection of underground utilities and evaluation of existing structures for support equipment during transportation to the pick site.

Tandem Picks: In certain situations, multiple cranes are required to make the necessary picks. Multiple crane, or tandem picks, could be advantageous for spans with limited access, or when hoisting large elements (on the basis of weight or size). Hoisting long bridge girders with a single crane may require complex rigging and multiple pick points, whereas using two cranes can simplify the required rigging and allow the operators better control over the load. It should be noted that a tandem pick is often considered a "critical lift" owing to the added coordination required among crane operators. Additional evaluation is required for critical picks during the planning phase.

Maximum Working Radius/Capacity: The crane must have the capacity for the entire hoisting operation; picking the element, swinging,

Figure 3-13. Demolition in a rural environment over an existing river. Source: Courtesy of Tighe & Bond.

articulating, and setting the element down. All stages of the operation should be considered when planning the pick.

Mobilization: The effort required for mobilizing and erecting a crane varies depending on the type of crane. Small AT cranes require very little effort to mobilize and arrive on site ready to pick. Larger AT cranes require counterweight to be mobilized separately. Rough terrain cranes have limited travel speed and often require transportation by other means. Large crawler cranes are mobilized as individual elements that require on-site assembly. Assembly can require multiple days and a support crane to erect.

Allowable Working Hours: The allowable working hours and overall time frame for crane picks are also an important consideration. Typically, demolition projects are in the vicinity of the public, railways, or waterways and may have limited time windows to operate. In these cases, a short setup and removal is imperative.

3.2.3 Rigging for Demolition

Rigging is required to secure hoisted elements and connect them to the crane's hoisting block. Typical rigging elements include slings, spreader beams and bars, and connecting elements. The size and type of rigging is dependent on the weight of the hoisted element, size of the element, and methods of connecting to the element.

Rigging for demolition or removal of members is different from rigging new material owing to the unknowns and variability that come with existing structures. Calculating the exact weight and center of gravity (CG) can be challenging, depending on the quality and availability of as-built drawings, undocumented repairs or overlays, build-up of debris, or other unanticipated conditions.

3.2.3.1 Rigging Elements

Slings: Slings are often used to connect the crane's hoisting block to the hoisted element. Multiple independent sling legs may be required to connect the hoisted element to the crane, but a minimum of two should be provided to ensure stability. Slings are available in various materials such as wire rope, nylon, or chain. The type of sling material is important to consider as each of them offers advantages and disadvantages that may prove critical to the operation.

Spreader Beams and Bars (Lifting Beams): These are engineered structural members located between the crane block and the lifted element, which is used to space out lifting points. Spreaders can also be used to limit additional loading in the lifted elements, owing to inclined rigging elements. Spreaders should be designed, tested, and permanently marked according to ASME *Design of Below-the-Hook Lifting Devices* (ASME 2023).

Connecting Elements: These are devices used to connect the elements of rigging to one another. Examples of these elements include shackles, turnbuckles, hooks, and padeyes. These devices can be used to simply connect elements or can also be used to add adjustments to the rigging system. Numerous devices are available that can be used in a rigging system. It is important to confirm that each device has been designed for the intended use and is appropriately sized for the application.

3.2.3.2 Rigging Design. The minimum rated capacity of rigging members should be clearly stated on the demolition plans. Several different factors should be considered when developing a rigging system. These factors include

- Center of gravity (CG) of picked element: Equilibrium of hoisted elements is found when the CG of a hoisted element is located directly below the hook of the crane. If the rigging is not designed to place the CG in this location prior to hoisting the element, the element will swing or translate to achieve equilibrium. The CG should be determined and the rigging designed to provide equilibrium to maintain stability of the hoisted element. Because of the approximate nature of CG determination for existing structures, workers should be mindful that the load may shift once freed.
- Total weight of pick: Cranes have a finite hoisting capacity. Loads more than this can lead to instabilities or structural failures. As such, it is critical to know the anticipated weight prior to hoisting the load. Additional information regarding the calculation of hoisted pieces can be found in Chapter 4.
- Lifting locations/pick points: It is important to carefully consider the locations from where the hoisted element will be hoisted. These locations affect the position of the CG in relation to the crane hook and directly relate to the structural demand on the picked element.
- Structural adequacy of picked element: The stability and structural adequacy of the picked element should be ensured for all stages of hoisting. When hoisting an element, there are often unique loading configurations that should be checked. These loading configurations can include high stress concentrations at pick points, stability concerns, and unusual demands.
- Attachment to picked element: It is important to ensure that the hoisted element is adequately secured to the rigging at all times. Various rigging configurations as well as multiple attachment types exist that can be utilized in the rigging system. When evaluating the attachments, attention should be given to stability, capacity, and constructability.
- Sharp corners and abrasive surfaces: During demolition, it is common for hoisted elements to have sharp corner and/or abrasive surfaces.

These should be considered when determining the appropriate rigging, as they can lead to damage and result in reduction in capacity or even failure of elements.

3.2.4 Typical Support Equipment

An often overlooked but essential part of planning a demolition project is the need for support equipment. Support equipment is required for erecting temporary supports, material management, personnel access, and equipment maintenance. Some common examples of support equipment are described in the following:

Water Trucks/Sprayers: Commonly used for dust suppression. Dust is a particular concern in urban environments.

Haul Trucks: Often used to transport and remove demolition debris from the construction site. Many trucks fall within the normal AASHTO live load provisions but should still be checked in combination with demolition equipment. Heavy haul trucks or earth/material movers often exceed AASHTO specifications (AASHTO 2020). These vehicles typically have short wheel bases and heavy axle weights, similar to the sample truck shown in Figure 3-14.

Figure 3-14. Specialty haul truck loaded with a crane.

Front-End Loaders: Commonly used to transport demolition debris or auxiliary equipment around the job site, as shown in Figure 3-15. Front loaders can be equipped with grapples, buckets, and fork attachments.

Skid Steers: Lighter equipment that serves a similar purpose to front-end loaders. Skid steers can either be wheeled as shown in Figure 3-16 or tracked.

Asphalt Milling Machines: Specialized equipment used to remove the asphalt overlay on bridge decks (Figure 3-17). These machines can also mill concrete to a limited depth. Hauling truck are often required in conjunction with milling machines to catch and transport debris.

Figure 3-15. Excavator with hammer attachment working with a front-end loader.

Figure 3-16. Skid steer used for moving material.

Figure 3-17. Milling machine used to remove asphalt/overlay prior to structural deck removal.

Personnel Access: Equipment that includes manlifts, scissor lifts, and snooper trucks. Personnel access equipment come in a wide range of sizes based on the lift height and articulation that is provided. Equipment size can be smaller than AASHTO design trucks but should be considered in relation to all other equipment.

3.3 CUTTING EQUIPMENT

Bridge demolition may require the cutting of bridge elements into manageable components that equipment can handle. Cutting is a time-consuming process that often requires specialized equipment and/or labor. Cutting structural members can change the fundamental behavior of structural elements and alter existing load paths. It is the responsibility of the engineer to understand the impacts on the behaviors, deformations, and capacities of structural elements that participate in a system being cut for removal. Cut locations on main structural members and bracing members should be analyzed to ensure that the remaining structure has adequate capacity and stability for the entire operation. The location and staging for cutting structural members should be clearly stated in the demolition plans, along with any restrictions on members being cut. Member restrictions may include, but are not limited to, assumed exposure duration for cut members, areas excluded for personnel or material storage, restrictions on the weight and type of equipment that may operate in the area, or minimum temporary bracing requirements.

3.3.1 Concrete Cutting

Concrete cutting is a specialized field and requires specific equipment for each type of cutting operation. When cutting concrete, it is important to carefully evaluate both the cut element and the remaining structure in relation to the cutting equipment used. For most applications, water is used to cool the concrete blades and protect workers from silica dust. Slurry water must be contained to prevent it from entering waterways or create slick work surfaces.

Slab Saw: Walk-behind cutting tool with a vertical, circular blade, typically used for cutting horizontal concrete members. Slab saws are often used in conjunction with panelized removal of bridge decks and barriers (Figure 3-18).

Concrete Core Hole Saw: large drill that cuts a circular hole into concrete using a core bit. For demolition operations, coring is typically used to provide access for rigging and to provide connections for structural systems. It is important to note that core drilling will also cut through reinforcing steel. Core bit sizes are commonly available in diameters ranging from 2 in. up to 24 in. (51 mm up to 610 mm).

Wall Saw: Wall-mounted saw that uses a rail to perform cuts in vertical elements (Figure 3-19).

Figure 3-18. Slab saw precutting sections of a concrete deck.

Figure 3-19. Wall saw cutting an existing retaining wall.

Wire Saw: Cutting system that uses a diamond-coated steel wire loop and a series of pulleys to cut through concrete. The saw system is mounted on the element being cut, with the cutting wire wrapped around the element being cut. Wire saws are typically used for large or oddly shaped elements, or where other cutting systems will not work (Figure 3-20).

Figure 3-20. Wire saw.

3.3.2 Steel Cutting

Steel cutting is typically performed using excavators equipped with shearing attachments or using handheld tools.

Excavator with Shears: Excavator-attached shears can be used to cut steel members and concrete reinforcing (see Section 3.2.1.2 for more information).

Demo Saw: Handheld saws that are typically used for cutting rebar.

Handheld Grinders or Shears: Lightweight tools typically used for thinner members or detailed cutting.

Rivet Busters: Hand-held tools typically used to remove steel rivets. These are often used for modification of existing steel structures during demolition operations.

Oxygen–Acetylene Torch: Hand-held torch that uses oxygen and acetylene to provide a high heat flame [between 2,462 °F (1350 °C) and 2,786 °F (1530 °C)] capable of cutting through steel elements. The flame preheats the steel to its melting point and gradually "cuts" through the member. Oxygen–acetylene torches are typically used for smaller members because of a slower production rate cutting through thicker steel elements.

Thermal Lance (also known as a burning bar or thermic lance): Thermal lance rods are a specialty cutting system that uses an iron or mild steel outer tube with internal iron, magnesium, or aluminum wires. The lance rod is hooked to a torch and an oxygen gas supply. Ignition is initiated with an oxygen–acetylene torch and reaches extremely high temperatures [between 7,000 °F (3871 °C) and 8,000 °F (4427 °C)]. Because of the high amount of heat that they generate, lances are capable of cutting through thicker steel members. (Figure 3-21).

Cutting through steel members typically produces sparks, which can present a danger of fire and should therefore be performed with caution. Most projects in higher fire danger areas will require a fire management plan and fire suppression equipment on site. Projects located in certain areas such as drought-stricken environments, near public traffic, or near rail tracks may also have prohibitions regarding the use of flame cutting.

3.3.3 Pile Cutters

A pile cutter is a manufactured apparatus comprised of hydraulic cylinders, a structural steel frame, and steel shears. The device is best suited to cut concrete piles or columns up to approximately 48 in. (1219 mm) in diameter. The system is able to operate above and below water.

Figure 3-21. Cutting an existing truss span with a thermal lance.

Pile cutters require the use of a crane, which provides support and hoists the system into place. Separate hydraulic pumps and lines are also required to power the cutter. Although suspended from the crane line, the pile cutter envelopes the pipe or column and using the hydraulic shear cuts through the member. During a cutting operation, the element being cut should be secured to a separate crane to maintain stability and to allow for removal after completion of cutting (Figure 3-22).

3.4 PROTECTION/MATERIAL MANAGEMENT SYSTEMS

Debris protection systems are often needed on bridge demolition projects to prevent debris from falling or migrating into areas outside the demolition work zone. Protective systems may also be required to prevent damage to structures that will remain during and after demolition. Project-specific specifications or requirements should be reviewed and followed, as they may specify requirements and design criteria for protection systems. The protection of the public, existing infrastructure, and new infrastructure is a critical component of a demolition project, and as such, protective systems may be integral to the project's success. Protective

Figure 3-22. Pile cutter on an existing concrete substructure.

systems are commonly required when demolition projects are near or over the following:

- Bridges and buildings,
- Waterways,
- Environmentally sensitive areas,
- Railways,
- Utilities, and
- Roadways, pathways, and so on.

The main function of protective systems for demolition projects is to prevent debris from exiting the project envelope where it could damage or contaminate the surrounding areas. A protective system is any device, structure, or component that can adequately provide this protection. Examples of typical protection systems are described in the following sections.

3.4.1 Modular Barges

Modular barges are commonly used as protection systems when working on bridges over waterways. Barges come in various sizes and may be modular connected or interlocked to create surface areas sufficient to meet the required protection area or stand alone units (Figure 3-23). If material is permitted to fall directly onto the barge, additional systems

Figure 3-23. Barges below excavators completing deck demolition.

may be required to prevent damage to the barge, such as wood mats or sand beds. Additional systems may also be required to prevent material from exiting the barge. As an added benefit, barges can serve as a working platform for demolition equipment such as excavators, cranes, and personnel access systems.

3.4.2 Fencing and Netting

Fencing and netting may be used as a shield to prevent demolition debris from leaving the work area. This type of protection is, in general, used to catch larger pieces of material and will not contain dust or smaller debris pieces. Fencing and netting is often provided only in specific locations and is moved as the demolition progresses (Figure 3-24). When using this type of protection, additional consideration should be given to the self-weight, wind loads, environment (e.g., de-icing), and debris loads for the design of the system and impacts on the existing structure.

3.4.3 Protective Shielding

Protective shielding is typically used to collect minor debris during the panelized removal of the concrete deck. As an added benefit, shielding can serve as access for personnel during demolition.

Protective shielding is typically comprised of timber lagging, plywood decking, or metal decking installed between the bottom flanges of existing steel bridge beams (Figure 3-25). Debris is typically removed by laborers or lifted out with demolition equipment using preplaced debris tarps.

Figure 3-24. Chain-link fence/net protection suspended by a crane. Source: Priestly Demolition.

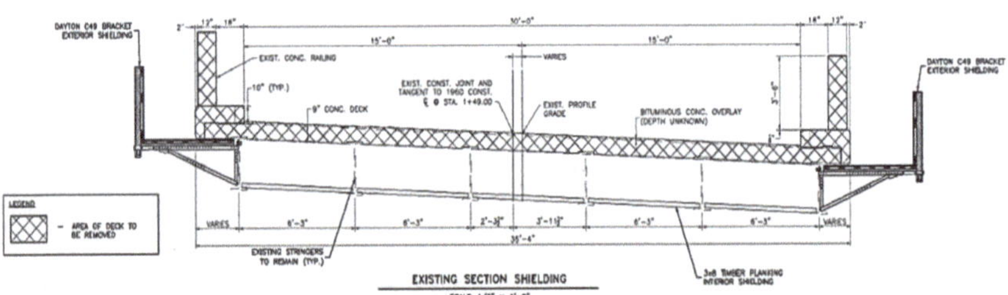

Figure 3-25. Typical protective shielding with timber lagging and bridge brackets.

Overhang brackets may also be used for shielding outside the exterior bridge beams (Figure 3-25).

3.4.4 Railroad Track Protection

Demolition of bridges near or over railways often requires protective systems to prevent material from damaging or fouling the tracks. Protection may be provided as a system that allows for the passage of rail traffic or as a shield placed directly over the tracks (Figure 3-26). Railroads often have specific requirements for the type of protection that is required for their tracks, and the railways should be consulted before determining the appropriate system. Railroads often require track protection to be detailed in the demolition plans and the plans must be submitted for approval prior to the beginning of any work.

Systems that allow for the passage of rail traffic are often independent structures that require design by a professional engineer. These systems should be designed for debris loading as well as any other relevant loads such as environmental or impact. These systems also require specified clearances that must be maintained for the passage of trains.

Shielding placed directly over tracks may be used only if the tracks are closed to rail traffic. It is often the case that tracks are closed only for limited windows, and as such, shielding systems require quick installation,

Figure 3-26. Sample railroad track protection structure.

clean-up, and removal. Shields typically consist of a solid surface located above the tracks to prevent damage to the rails from falling debris. A damping material to *soften* the fall such as rubber or fill may also be added to the shield. It is common for ballasted track to be lined with geofabric to protect the ballast from contamination.

3.5 TEMPORARY BRACING/SHORING SYSTEMS

Certain bridge demolition projects require the use of additional, external supports, bracing, and / or temporary members. A structural analysis of the structure during all phases of the bridge demolition operation is typically required to determine the need for additional support members. This analysis should be performed by a registered professional engineer with experience in bridge demolition. Typical support systems include

- External support towers,
- Existing member bracing,
- Temporary structural members needed to maintain load paths,
- Bearing reinforcement and / or lock-up, and
- Existing member strengthening.

Design of temporary bracing / shoring should be done in accordance with applicable codes, such as AASHTO's *Guide Design Specifications for Bridge Temporary Works* (AASHTO 2017a) for vehicular bridges, or other relevant codes or owner design specifications.

3.6 HEAVY LIFT EQUIPMENT

The following section includes an overview of specialty equipment commonly used for removal of large sections or sometimes entire bridge spans units. Because of high cost and highly technical requirements, these types of operations are typically used only when the sequential removal of the structure is impractical, or when the schedule does not allow for traditional removal methods. When this method is used, large bridge sections are removed and processed off-site.

3.6.1 Strand Jacks

Strand jacks are hydraulic jacks that utilize steel strands to raise or lower elements. Strand jacks often have large capacities, which make them effective tools for the removal of large bridge sections. A typical application for strand jacks is to move full truss or girder spans in the vertical or horizontal direction (Figures 3-27 and 3-28).

Figure 3-27. Strand jack lowering an old simple span railroad truss. Lifting lugs were bolted to an existing structure.
Source: (Top) Courtesy of Walsh Construction Company.

Extensive engineering and coordination are required to safely execute operations using strand jacks. A particularly challenging element is the design of a connection between the strand jack anchor block and the existing structure (Figure 3-28). Often, the engineer and contractor develop a solution that requires temporary strengthening and, in some cases, additional members to distribute highly localized reactions.

Figure 3-28. Span attachment strand jack lowering a truss span with an underslung beam.

Another challenge is determining an adequate location for placing the strand jack units and wire strands. Strand jacks are commonly mounted atop temporary falsework towers or temporary platform brackets secured to the remaining elevated structure. Careful planning is required when establishing the size of the work platforms for strand jack operations. In addition to the high-capacity jacks, a large volume of other miscellaneous specialized equipment is required (Figure 3-29).

Figure 3-29. Strand jacks and recoiler units mounted above to an existing structure.

3.6.2 Self-Propelled Modular Transporters

Self-propelled modular transporters (SPMTs) are often used to remove or install complete bridges for rapid replacement. SPMTs can be used in a variety of projects, including those that require speedy changeouts, projects where temporary shoring is not practical, or where traditional demolition methods using excavators and cranes are not feasible. SPMT units often have large load-carrying capacity, and the total capacity of the system can be increased by adding additional units/axels. The ability to handle large loads may reduce preparatory work before removal.

There are restrictions for using SPMTs. The travel path between the staging area and the bridge site must have sufficient capacity and be relatively flat and clear for the SPMT to operate. This may require temporary improvements or site modifications for access. Overhead clearances must be checked for the entire system. Local utilities must be identified and protected if necessary.

Typically, SPMT units are equipped with a fixed tower system or stacks of grillage that transfer the load from the bridge into the units (Figure 3-30). Some projects require SPMTs outfitted with falsework tower systems furnished with adjustable jacking towers to raise/lower a bridge. Additional jacks within the SPMTs can also be used for the raising and lowering operations. These internal jacks can also be used to adjust the axle heights to accommodate for grade changes during transport. SPMTs typically have jack strokes of around $\pm$ 12 in. (305 mm) (24 in. (610 mm) total) and have minimum collapsed heights between 42 in. (1067 mm) and 48 in. (1219 mm).

Figure 3-30. Self-propelled modular transporter removal of an existing railroad structure over an existing roadway.

The following checklist should be considered when determining whether SPMTs are suitable for a specific bridge removal project.

1. Site constraints and contours along the travel path:
   - Similar to tracked cranes, SPMTs can travel with load only at certain grade changes. The terrain along the travel path should be considered, and regrading may be required.
   - Temporary access ramps may be required if the area under the element to be supported is not easily accessed.
   - Soil stabilization along the travel path may be required because of large axel loads.
   - Loading above underground utilities must be considered. Provide protection to utilities where required.
2. Condition of the existing structure and alignment of support points with SPMTs: Existing structure may require local strengthening at support locations. This is more critical if SPMT support points occur at interior locations, away from existing structure supports.
3. Project budget and schedule:
   - Typically, SPMT moves are more expensive than traditional demolition. However, this is dependent on numerous variables associated with traditional demolition methods, including the size of equipment needed to perform the removal operations, the need for temporary supports, traffic control/closures, and schedule impacts.
   - In urban areas or on railroad projects where long-term closures are not allowed, the schedule requirements may offset the cost.

Refer to the FHWA *Manual on Use of Self-Propelled Modular Transports to Remove and Replace Bridges* (FHWA 2007) for more information on bridge removal with SPMTs

3.6.3 Span Float-Out Barges

Barges are used for heavy bridge demolition projects where operations occur over water. Barges and SPMTs or strand jacks may be used in combination if lowering is required over water, or if unloading the existing structure is required after float-out (Figure 3-31).

Barges may also be used to lift existing bridge elements without the aid of strand jacks or SPMTs. The loading to the barges is similar, but the means with which the structural elements are lifted varies. Standard methods of lifting elements include the following:

1. Jacking towers with large adjustable strokes mounted on barges.
2. Pumping of ballast water. For this method, barges come in ballasted with water and fixed height towers on the deck; once the barge is in position, ballast water is pumped out, which swaps the ballast load for the bridge span weight.

Figure 3-31. Self-propelled modular transporter supporting an existing bascule span for float-out and then unloading onto shore for processing.

3. Natural tide cycles. For bridges in tidal regions, barges can be used to remove spans by positioning the barge under the span at low tide and then using high tide to raise the barge and lift the bridge. Careful planning is required to verify whether the anticipated change in tide is greater than the anticipated draft once the barge is loaded with the bridge weight. Because of the tide cycle, this operation imposes time constraints on the operation.

The evaluation of barges for bridge float-outs is not standardized and requires project-specific evaluation. The American Bureau of Shipping (ABS) provides barge design standards; however, these design guidelines are intended for uniform barge loading for material transport. Barges often require cribbing or falsework to support the bridge structure and adequately distribute the load to the barge. Cribbing or falsework loading may require the following: additional layers of grillage and/or leveling medium used to spread out the load to a larger area of the barge deck, or securing grillage supports to the side hulls or other stiffened elements of the barges. Barge structural conditions can vary widely, on the basis of size, original design use intent, years of service, maintenance history, and exposure to salt water. An inspection by a qualified marine engineer or naval architect performing the barge float-out analysis is recommended. The condition of the barge's internal framing at the load entry points is a critical design aspect. Structural evaluations must include global bending (hogging and sagging moments) and local analysis at load entry points. In some cases, internal barge strengthening is required to handle local load entry demands. The barge demands from the bridge removal should also be checked with all other loads present on the barge, including barge equipment, demolition equipment, materials, and any other loads necessary to complete the work See Figures 3-32 and 3-33 for sample float-out configurations.

Figure 3-32. Existing truss after a span lowered with strand jacks supported on steel towers.

Figure 3-33. Existing bascule leaf supported on timber cribbing.

In addition to barge cribbing supports, the decks of barges used for float-outs are commonly filled with a variety of other equipment. Some of the common equipment includes

- Connex boxes (i.e., breakrooms/tool storage),
- Manlifts,
- Ballast pumps (heated covers if needed),
- Winches and mooring lines,
- Light plants, and
- Portable toilets and washrooms.

REFERENCES

AASHTO (American Association of State Highway and Transportation Officials). 2017a. *Guide design specifications for bridge temporary works*. 2nd ed. Washington, DC: AASHTO.

AASHTO. 2020. *LRFD bridge design specifications*. 9th ed. Washington, DC: AASHTO.

ASME (American Society of Mechanical Engineers). 2019. *Planning for load handling activities*. P30.1. New York: ASME.

ASME. 2023. *Design of below-the-hook lifting devices*. BTH-1. New York: ASME.

FHWA (Federal Highway Administration). 2007. *Manual on use of self-propelled modular transporters to remove and replace bridges*. Washington, DC: FHWA.

CHAPTER 4
LOADS

Bridge demolition analysis is similar to erection analysis in that both must consider the effects from dead, live, and environmental loading. Dead loads include all permanent components of the structure, such as primary and secondary load-carrying members, decks and overlays, barriers, railings, and ancillary structures. Live loads include traffic, pedestrians, construction personnel, equipment, shielding, stockpiles, and all other transient loading that may be on the structure during the demolition process. Environmental loading is usually controlled by wind, especially as the structure is partially deconstructed, but loading such as ice, snow, and seismic events should be considered on a project-specific basis. The controlling load combinations should be checked during the progressive steps of demolition to ensure strength and stability of the partially deconstructed bridge.

Although modern bridge design is typically performed using load and resistance factor design (LRFD), allowable stress design (ASD) is often preferred for bridge demolition because unfactored loads are required for calculating deflections and determining pick weights and pick locations. More information regarding the use of ASD versus LRFD is presented in Chapter 2. It is the responsibility of the engineer to determine the most appropriate design philosophy, load cases, load combinations, and load factors (if necessary) for each demolition project that meets project-specific requirements.

4.1 DEAD LOADS

4.1.1 Introduction

It is important to establish an accurate estimate of the structure self-weight in its current condition to perform a realistic demolition analysis.

This typically involves reviewing the original design drawings, as well as any retrofit or redecking plans and inspection reports that may be available to gain an understanding of the full history of the structure.

In some cases, the existing plans will provide either reaction or member force information that may be used to validate the calculated dead load used for demolition analysis. Field verification is recommended, when possible, to confirm material properties and existing conditions.

4.1.2 Determination of Dead Load

The best way to determine the self-weight and center of gravity of an existing structure is by performing a weight takeoff using existing bridge design documents. The level of detail used in the takeoff will depend on the structure type, condition of the structure, project timeline, removal methods, and availability of bridge design documents. Engineering judgment is required to determine the appropriate level of detail required. Minimum unit weights of materials should be in accordance with AASHTO's *LRFD Bridge Design Specifications*, Table 3.5.1-1 (AASHTO 2020), in the absence of more precise information provided in the existing plans or specifications.

The level of detail used for a weight takeoff can vary dramatically. At the preliminary stages of design, a rough takeoff may include weights only for the main load-carrying members of the structure, such as the deck, girders, or truss members. At the high end of effort, a thorough takeoff is performed, in which the weights of all bridge elements are considered.

In all cases, the use of a detail factor is recommended to adjust the weight according to the level of detail used and account for undocumented changes such as retrofits or debris buildup. The detail factor is a modifier that is typically applied as a percent increase on the main members that were directly calculated. A rough weight takeoff would typically necessitate applying a more substantial detail factor compared to a more thorough weight takeoff.

A list of common structure types and minimum detail factors to be added to the main member weight is provided in the following paragraph. Note that the detail factors listed are recommended only for preliminary design and should be confirmed either by dead load reaction information provided in the plans (if available) or by performing a more detailed takeoff of a few representative members. Certain structures (i.e., riveted structures) may require detail percentages in excess of 40%. Even with a detailed weight takeoff, a minimum 5% increase is recommended. Note that the detail percentages provided are intended to be applied to the main load-carrying members and do not include an allotment for deck, rail, or other miscellaneous nonstructural components:

- Steel girders 10%,
- Steel trusses 25%, and
- Steel arches 20%.

4.1.3 Lifted Loads

In general, it is recommended that a more detailed takeoff be performed for estimating the self-weight of lifted loads, which are used to size the removal equipment. It is important to remember that when a member to be removed is cut free, it often cannot be set back down. This makes estimating the pick weight and sizing the equipment crucial to safe removal operations. This is especially important for cases where multiple pieces of equipment are required for a given removal operation.

If a detailed weight takeoff is not feasible owing to a lack of available existing plans, then it is the engineer's responsibility to use alternate methods to determine a design pick weight. Without plan information, a detailed site inspection with field measurements should be performed. In addition, removal of the structure with methods that avoid hoisting large sections and decreasing the allowable equipment capacity should be considered.

Removal operations also introduce the possibility for members to get bound up or require additional force to free them from the remaining structure. The FHWA publication *Engineering for Structural Stability in Bridge Construction* (FHWA 2015) suggests considering a minimum dynamic load allowance equal to 20% of the calculated member weight to account for lifting resistance caused by interaction with other members, miscellaneous debris on the member, and so on.

In addition, many cranes are equipped with load moment indicators (LMIs), which provide a readout of the load on the hook, and thus offer some degree of confirmation of the estimated pick weight during lifting. This can be monitored on initial lifts to confirm loads and gain confidence in anticipated pick weights for future or more critical lifts.

4.2 GENERAL LIVE LOADS

4.2.1 Introduction

Live loads to be considered during bridge demolition typically include equipment, personnel, construction material, debris, and debris shields. It is common practice to compartmentalize these loads as localized equipment loads and distributed live loads. Construction equipment loading is discussed separately in Section 4.3 because of the unique considerations and magnitudes of loads.

Construction live loads are typically dependent on the removal process used and can often be restricted by the engineer. These loads should be applied and staged in the analysis to be representative of how the work will be performed. All construction live load assumptions should be clearly noted in the plans, including but not limited to areas of operation, maximum anticipated loads, and areas of exclusion.

4.2.2 Uniform Distributed Loads

Uniform live load during demolition typically includes weight from construction personnel, as well as tools and materials stored on the bridge during demolition. ASCE 37-14, Table 4-4 (Figure 4-1) (ASCE 2014), has recommendations for uniform live load depending on the type of work being performed. This generic uniform live load can vary or be eliminated entirely based on the engineer's discretion and noted restrictions in personnel/materials/equipment.

Table 4-4. Classes of Working Surfaces for Combined Uniformly Distributed Loads

Operational Class	Uniform Load[a] [psf (kN/m^2)]
Very Light Duty: sparsely populated with personnel, hand tools, very small amounts of construction materials.	20 (0.96)
[b]Light Duty: sparsely populated with personnel, hand-operated equipment, staging of materials for lightweight construction.	25 (1.20)
[b]Medium Duty: concentrations of personnel, staging of materials for average construction.	50 (2.40)
[b]Heavy Duty: material placement by motorized buggies, staging of materials for heavy construction.	75 (3.59)

[a]Loads do not include dead load, D, construction dead load, C_b, or fixed material loads, C_{FML}.

[b]OSHA categories.

Figure 4-1. ASCE 37-14, Table 4-4.
Source: ASCE (2014).

4.2.3 Material Storage and Shielding

It is common for debris material to be stored on the bridge during demolition activities until it can be hauled off for disposal. It is important to consider both the magnitude and the location of any stockpiled materials when investigating the effects on the overall structure. Any restrictions should be noted in the demolition plan.

The contractor is often required to install protective shielding to prevent debris from falling onto railroads, active roadways, other infrastructure, waterways, or environmentally sensitive areas (see Section 3.4 for various shielding and material management systems). Protective shielding adds weight on the structure resulting from the self-weight of the shielding and collected debris along with its dynamic impact. Note that during demolition, structural elements considered in the self-weight loading may become debris loading and do not need to be double-counted (i.e., debris generated by hammering to remove the concrete deck does not need to be considered on top of the full concrete deck weight). Although removal of deck concrete does not increase self-weight, the capacity of the bridge will likely be reduced owing to a loss of composite section.

4.3 REMOVAL EQUIPMENT LOADS

4.3.1 Introduction

During demolition, contractors have a wide variety of equipment at their disposal to aid in various removal methods. Refer to Chapter 3 for details describing various pieces of equipment and what activities they are most useful for.

Bridges must be analyzed to verify structural adequacy at all stages of demolition for the loading, location, and dynamic effects of all major pieces of equipment planned for use. AASHTO's *Guide Design Specifications for Bridge Temporary Works*, Section 2.3.4.5 (AASHTO 2017a), provides guidance for determining loading for both rated and unrated equipment. A given piece of equipment may come in various models and configurations, and thus, it is important to consider the actual equipment configuration planned for use when developing the associated loads on the structure.

4.3.2 Excavators

4.3.2.1 Overview. Excavators are mounted either on tracks or wheels. The manufacturer specifications often report ground-bearing data as the total weight of the machine divided by the total ground contact area. This does not accurately reflect the load distribution to the tracks or wheels during demolition operations. It is the engineer's responsibility to determine this loading based on the machine weight, attachment, and operating radius.

Currently, there is no publicly available software to compute bearing pressures for excavators. Some excavator manufacturers will provide track loading data for specific load and reach scenarios. This results in an inefficient process if equipment, attachments, or reach change during the design and limits flexibility in the field during operations. Common industry practices have led to the development of methods used to approximate track pressures for excavators. A select few of these methods are provided.

4.3.2.2 Load Development. Developing load effects from excavators is complicated because of the various moving parts. The weight of the boom, stick, attachment, supported load, body, and counterweight all rotate as the machine swings, resulting in varying load distribution. In addition, as the excavator's arm is extended in and out, the center of gravity of the boom, stick, and supported load are all variable. Considering all these variables, a simplified method of developing track pressures is necessary.

The first step in developing load effects from an excavator during demolition is to collect necessary information on the machine and operation. A list of standard information required is detailed as follows:

1. Machine model number;
2. Working radius and operating limits: (a) Does the excavator need to work with a boom over the front or over the side? (b) Does the excavator need to swing or walk with load?
3. Overall machine weight;
4. Manufacturer provided lift charts (included in machine specifications);
5. Machine geometry and component weights (track gauge, width, contact length, boom, stick, counterweights);
6. Demolition-specific attachment weight (shear/hammer/breaker/specialty attachment);
7. Pick weight (if applicable); and
8. Demolition activity (used primarily to determine dynamic impact effects).

4.3.2.3 Track Pressure Development. The methods described for developing excavator track pressures treat the tracks like spread footings, similar to how crane manufacturers develop track pressures for crawler cranes. Track pressures are either trapezoidal or triangular if working over the front of the machine, or uniform but unequal if working over the side of the machine, as shown in Figure 4-2.

As noted previously, a simple calculation of the total weight divided by the total track area will result in unconservative bearing pressures, which is the reason that this method is not recommended.

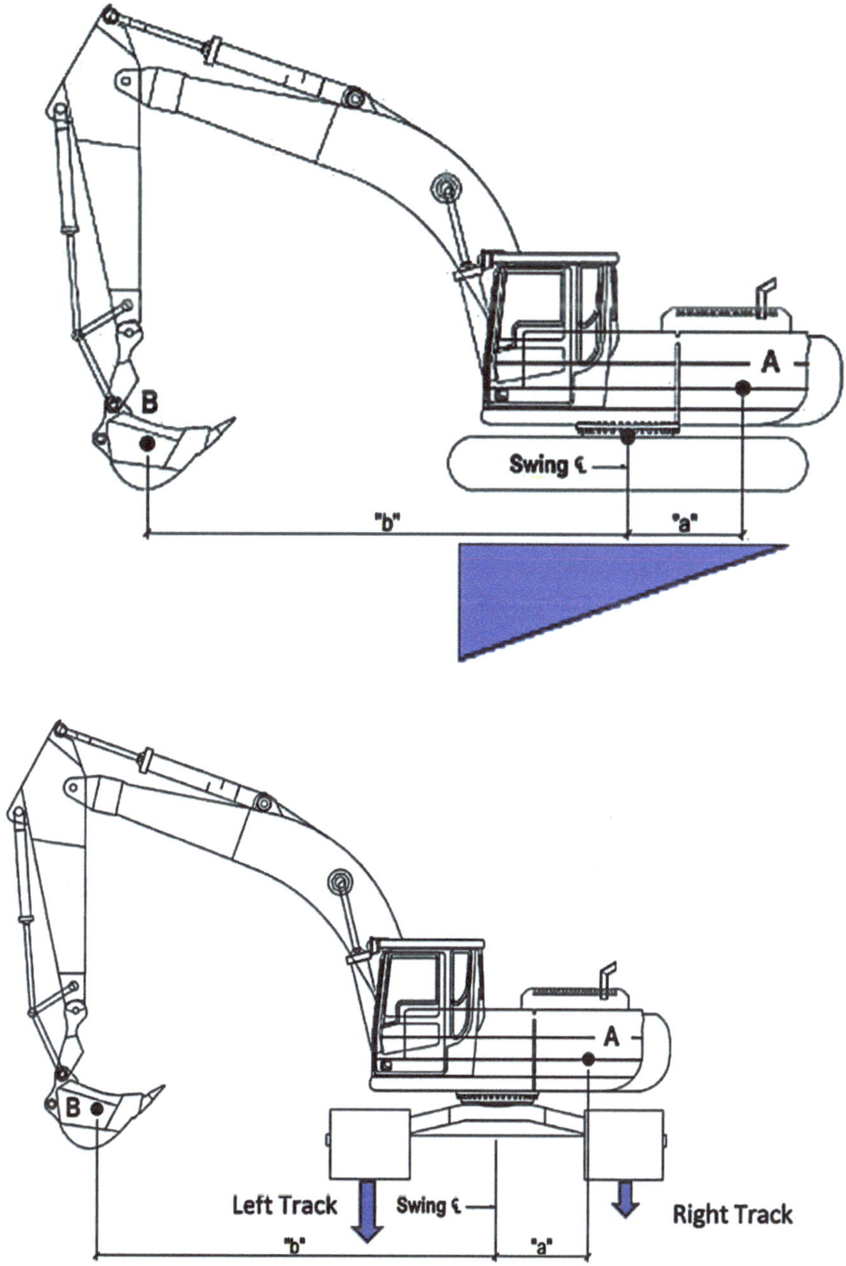

Figure 4-2. Schematic of track pressures for excavator working over the front and over the side.

Details of the three most common methods for determining track pressure are discussed.

1. Simplified Method: An approximate method of estimating excavator track pressures is by assuming a common uniform distribution of the total weight divided between tracks. This method should be used only for order-of-magnitude studies (i.e., feasibility studies or preliminary analysis) where the goal is to gain a quick understanding of the equipment load distribution or when the exact machine information and/or operation limits are unknown. Based on typical industry practice, it is recommended to use a 60%/40% to 80%/20% split between tracks, depending on the working radius and pick weight. This method is not considered a substitute for a more detailed load development technique to be used for the final design.
2. Capacity Chart Method: A more refined method of estimating track pressures is to utilize the excavator capacity charts provided in the manufacturer specifications, where capacities controlled by 75% of the tipping load (the load that theoretically balances the excavator) are given. Using statics, the tipping moment is used to determine the approximate excavator center of gravity for the configuration noted in the chart. The approximate C.G. is then used to determine track pressures for the same excavator configuration with a different attachment or pick load (Figure 4-3). It is important to note that when using this method, the excavator capacity taken from the chart must be controlled by tipping, rather than the hydraulic capacity of the machine.
3. Detailed Component Method: Another method for estimating track pressures is to develop a spreadsheet or FEA model that includes a detailed summary of all excavator parts, weights, and positions during a given operation. See Figure 4-4 for an example of this method. For guidance on developing track pressures using this method, refer to *Cranes and Derricks* (Shapiro and Shapiro 2010), which provides information for similar applications for tracked cranes. Note that this method requires a detailed breakdown of various equipment part weights and therefore may not be deemed feasible or necessary.

Additional Resources: The Federal Piling Specialists (FPS) organization has developed a published spreadsheet tool for developing track pressures for a crane or a pile rig. The pressures are meant to be used for loading in the design of support platforms. A sample of the spreadsheet is shown in Figure 4-5. The principles are similar to those described in the aforementioned methods.

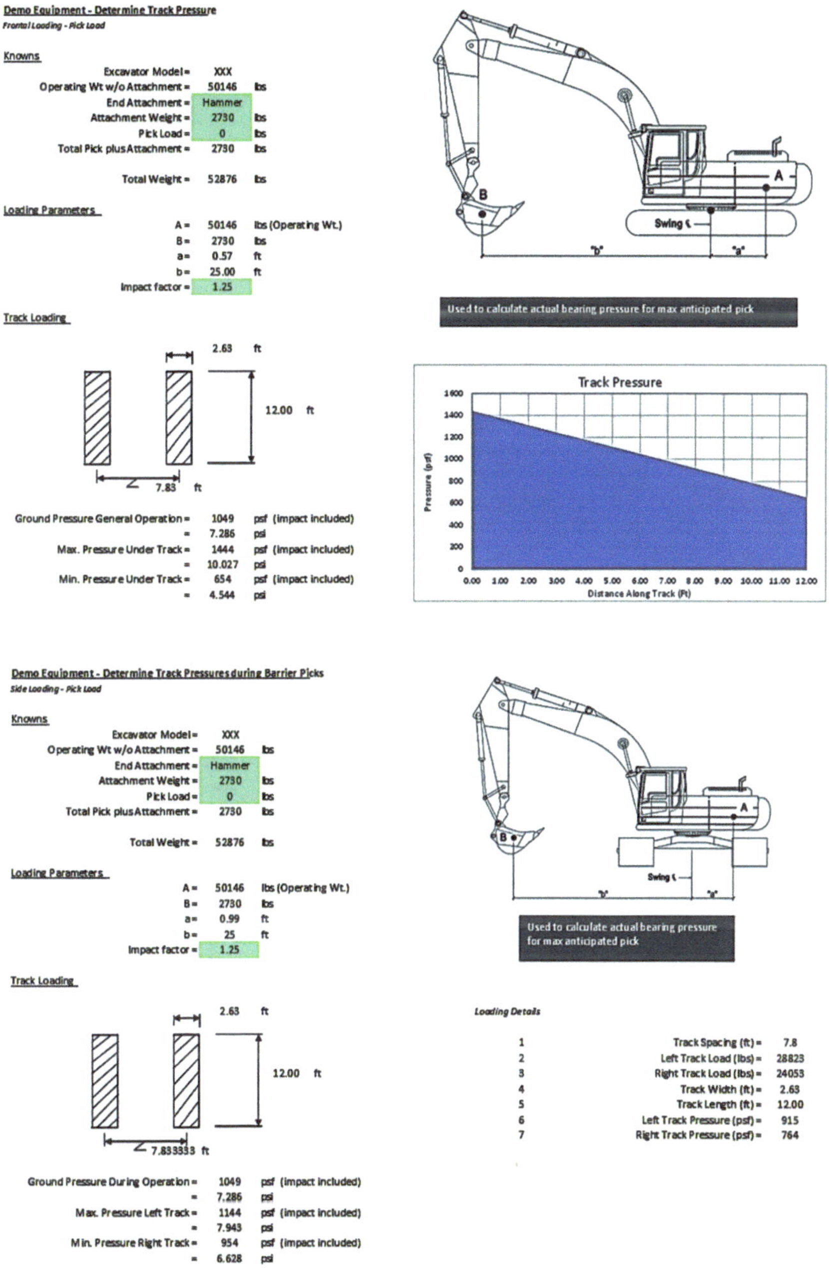

Figure 4-3. Sample output of a capacity chart method for developing excavator track pressures.

Note: This figure is an example only and does not reflect real data from an actual machine.

EXCAVATOR CRAWLER PRESSURE CALCULATION WITH BUCKET [3rd MEMBER]		
INPUTS		
Load Weight (kip)	W_{lo}	5.00
Attachment Weight (kip)	W_a	5.40
Distance From CG of Load to CL of Rotation (ft)	d_{lo}	30.00
Distance From CG of Attachment to CL of Rotation (ft)	d_a	30.00
Distance From CG of Boom to CL of Rotation (ft)	d_b	11.75
Distance From CG of Stick to CL of Rotation (ft)	d_s	26.50
MACHINE INFORMATION		
Boom Weight (kip)	W_b	10.20
Stick Weight (kip)	W_s	6.10
Upperstructure Weight (kip)	W_u	54.50
Carrier Weight (kip)	W_c	34.60
Distance From CG of UpperStructure to CL of Rotation (ft)	d_u	-5.53
Distance From CG of Carrier to CL of Bearing (ft)	d_c	0.00
Distance From CL of Rotation to CL of Bearing (ft)	x_0	0.00
Center to Center Distance of Crawlers (ft)	d_t	9.00
Crawler Bearing Width (ft)	w	2.92
Maximum Bearing Length (ft)	d_l	14.33
CALCULATIONS		
Net Moment Above Swing Circle (kip-ft)	M_u	292.12
Total Vertical Load (kip)	V	115.80

TABULATED RESULTS								Heavy Crawler		Light Crawler	
α (Deg)	M_{nf} (kip-ft)	M_{ns} (kip-ft)	R_h (kip)	R_l (kip)	e (ft)	l_b (ft)	Distribution Shape	q_{max} (ksf)	q_{min} (ksf)	q_{max} (ksf)	q_{min} (ksf)
0	292.1	0.0	57.9	57.9	2.52	13.93	Triangular	2.85	0.00	2.85	0.00
5	291.0	25.5	60.7	55.1	2.51	13.96	Triangular	2.98	0.00	2.70	0.00
10	287.7	50.7	63.5	52.3	2.48	14.04	Triangular	3.10	0.00	2.55	0.00
15	282.2	75.6	66.3	49.5	2.44	14.19	Triangular	3.20	0.00	2.39	0.00
20	274.5	99.9	69.0	46.8	2.37	14.33	Trapezoidal	3.29	0.01	2.23	0.01
25	264.7	123.5	71.6	44.2	2.29	14.33	Trapezoidal	3.35	0.07	2.07	0.05
30	253.0	146.1	74.1	41.7	2.18	14.33	Trapezoidal	3.39	0.15	1.91	0.08
35	239.3	167.6	76.5	39.3	2.07	14.33	Trapezoidal	3.41	0.25	1.75	0.13
40	223.8	187.8	78.8	37.0	1.93	14.33	Trapezoidal	3.41	0.36	1.60	0.17
45	206.6	206.6	80.9	34.9	1.78	14.33	Trapezoidal	3.38	0.49	1.46	0.21
50	187.8	223.8	82.8	33.0	1.62	14.33	Trapezoidal	3.32	0.64	1.33	0.25
55	167.6	239.3	84.5	31.3	1.45	14.33	Trapezoidal	3.24	0.80	1.20	0.29
60	146.1	253.0	86.0	29.8	1.26	14.33	Trapezoidal	3.14	0.97	1.09	0.34
65	123.5	264.7	87.3	28.5	1.07	14.33	Trapezoidal	3.02	1.16	0.98	0.38
70	99.9	274.5	88.4	27.4	0.86	14.33	Trapezoidal	2.88	1.35	0.89	0.42
75	75.6	282.2	89.3	26.5	0.65	14.33	Trapezoidal	2.72	1.55	0.81	0.46
80	50.7	287.7	89.9	25.9	0.44	14.33	Trapezoidal	2.54	1.75	0.73	0.51
85	25.5	291.0	90.2	25.6	0.22	14.33	Trapezoidal	2.35	1.96	0.67	0.55
90	0.0	292.1	90.4	25.4	0.00	14.33	Uniform	2.16	2.16	0.61	0.61

SUMMARY OF RESULTS FOR HEAVY CRAWLER				
Maximum Triangular Loading	q_{max}	3.20	q_{min}	0.00
Maximum Trapezoidal Loading	q_{max}	3.41	q_{min}	0.25
Maximum Uniform Loading	q_{max}	2.16	q_{min}	2.16
Maximum Triangular Distributed Loading (kip/ft)	Q_{max}	9.35	Q_{min}	0.00
Maximum Trapezoidal Distributed Loading (kip/ft)	Q_{max}	9.96	Q_{min}	0.72
Maximum Uniform Distributed Loading (kip/ft)	Q_{max}	6.31	Q_{min}	6.31

Figure 4-4. Sample output of a detailed component method for developing excavator track pressures.
Note: This figure is an example only and does not reflect real data from an actual machine.

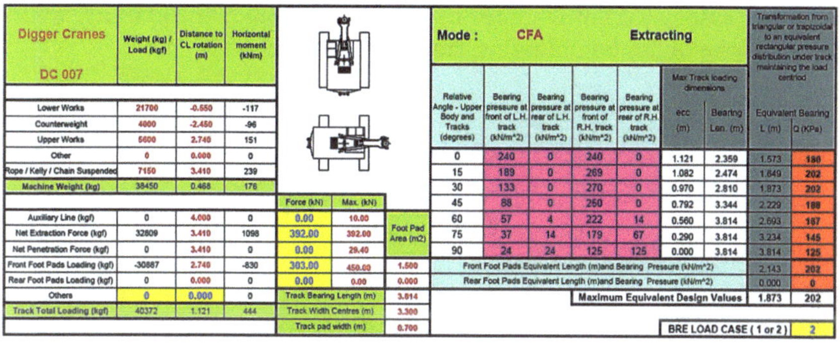

Digger Cranes DC 007	Weight (kg) / Load (kgf)	Distance to CL rotation (m)	Horizontal moment (kNm)
Lower Works	21700	-0.550	-117
Counterweight	4000	-2.450	-96
Upper Works	5600	2.740	151
Other	0	0.000	0
Rope / Kelly / Chain Suspended	7150	3.410	239
Machine Weight (kg)	38450	0.468	176
Auxiliary Line (kgf)	0	4.000	0
Net Extraction Force (kgf)	32809	3.410	1098
Net Penetration Force (kgf)	0	3.410	0
Front Foot Pads Loading (kgf)	-30887	2.740	-830
Rear Foot Pads Loading (kgf)	0	0.000	0
Others	0	0.000	0
Track Total Loading (kgf)	40372	1.121	444

	Force (kN)	Max. (kN)	Foot Pad Area (m2)
Auxiliary Line	0.00	10.00	
Net Extraction Force	392.00	392.00	
Net Penetration Force	0.00	29.40	
Front Foot Pads Loading	303.00	450.00	1.500
Rear Foot Pads Loading	0.00	0.00	0.000
Track Bearing Length (m)			3.814
Track Width Centres (m)			3.300
Track pad width (m)			0.700

Mode : CFA Extracting					Max Track loading dimensions		Transformation from triangular or trapizoidal to an equivalent rectangular pressure distribution under track maintaining the load centriod	
Relative Angle - Upper Body and Tracks (degrees)	Bearing pressure at front of L.H. track (kN/m^2)	Bearing pressure at rear of L.H. track (kN/m^2)	Bearing pressure at front of R.H. track (kN/m^2)	Bearing pressure at rear of R.H. track (kN/m^2)	ecc (m)	Bearing Len. (m)	Equivalent Bearing L (m)	Q (KPa)
0	240	0	240	0	1.121	2.359	1.573	180
15	189	0	269	0	1.082	2.474	1.649	202
30	133	0	270	0	0.970	2.810	1.873	202
45	88	0	250	0	0.792	3.344	2.229	188
60	57	4	222	14	0.560	3.814	2.693	187
75	37	14	179	67	0.290	3.814	3.234	145
90	24	24	125	125	0.000	3.814	3.814	125
Front Foot Pads Equivalent Length (m)and Bearing Pressure (kN/m^2)							2.143	202
Rear Foot Pads Equivalent Length (m)and Bearing Pressure (kN/m^2)							0.000	0
Maximum Equivalent Design Values							1.873	202
BRE LOAD CASE (1 or 2)								2

Figure 4-5. Sample output from Federal Piling Specialists ground-bearing pressure calculator for machines performing drilling operations.

4.3.3 Cranes

Crane loading can result either from a crane traveling over a structure or from a crane performing an actual pick from a given location while supported by the structure.

The loading induced by a given crane is dependent on the type of crane being used. Crawler cranes are mounted on tracks and, as a result, are similar to excavators in terms of the loading induced on the structure. Mobile, all-terrain, truck, and rough terrain (RT) cranes are mounted on wheels, but most often, they load the structure through outriggers when picking.

Development of a Bridge Construction Live Load Analysis Guide, developed by the Wisconsin Department of Transportation (2011), provides guidance related to crane loading. According to this document, the determination of crane load must consider the following variables:

- Make and model of the crane, including the amount and distribution of counterweights;
- Maximum lift, including rigging weights;
- Swing limits;
- Maximum radius of lift; and
- New construction versus demolition activities.

It is the engineer's responsibility to evaluate which potential crane load cases need to be evaluated. Some common cases that may control design include

- Maximum load, maximum radius;
- Minimum load, minimum radius; and
- No load, minimum radius.

Most crane manufacturers have software resources that can be used to develop ground-bearing pressures for tracked equipment or outrigger reactions for mobile units. The following is a list of crane manufacturers and their associated software for use in determining crane loading, which can either be obtained online for free or purchased:

- Liebherr,
- Crane Planner 2.0,
- LICCON,
- Manitowoc/Grove,
- Ground-Bearing Pressure Estimator,
- Outrigger Pad Load Calculator,
- Linkbelt,
- Ground-Bearing Pressure Calculator,
- Outrigger Pad Load Calculator,
- Terex/Demag,
- Terex Cranes Lift Plan,
- Cranimation,
- Kobelco,
- Ground-Bearing Pressure Calculator,
- Tadano, and
- Outrigger Load Calculator.

In addition, the computer program 3D Lift Plan (A1A Software 2023) can provide loading for a variety of crane makes and models. If loading information is not available, it may be calculated on the basis of the crane's weight and geometric data using statics, similar to the methods described previously for excavators.

Figures 4-6 and 4-7 show typical printouts for manufacturer-provided crane track pressures and outrigger loading, respectively. Note that when

Manitowoc

Ground Bearing Pressure Estimator

GBPressure ver: 1.0.6.9
5/24/2022 9:46 AM

Project: Test	Operating Surface: Hard - Tread Contact Width 20.3 Inches
Model: M-250 Series 3	Primary: 190FT (57.9M) #44 HL BOOM (US)
Machine Counterweight: 248,700-LB (112,810-KG) + 120,000-LB (54,430-KG)	Secondary: 0FT (0M) No Attachment
Crawler:Fixed, Position: Gantry Up	Load: From Boom 90,000 LBS @ 80 ft radius
Crane Total Weight Including Load: 753,030 LBS	

Imperial Pressure Conversion Factor: 1 PSI = 144 PSF

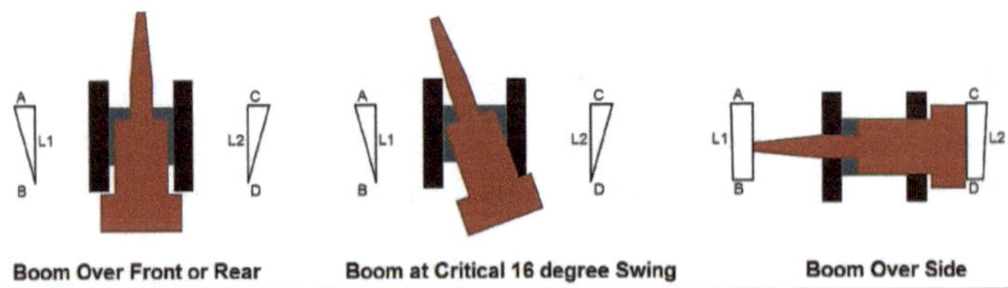

Boom Over Front or Rear		Boom at Critical 16 degree Swing	Boom Over Side
Center of Rotation to Fulcrum: 136.25 inch		Center of Rotation to Fulcrum: 136.25 inch	Center of Machine to Ctr of Crawlers: 132 inch
A	189.5 psi	196.2 psi	107.2 psi
B	0.0 psi	0.0 psi	100.0 psi
L1	196.3 inch	216.7 inch	272.5 inch
C	189.5 psi	169.9 psi	36.5 psi
D	0.0 psi	0.0 psi	29.3 psi
L2	196.3 inch	187.6 inch	272.5 inch

Figure 4-6. Ground-bearing pressure estimator printout for a crawler crane. Source: Courtesy of Manitowoc Cranes.

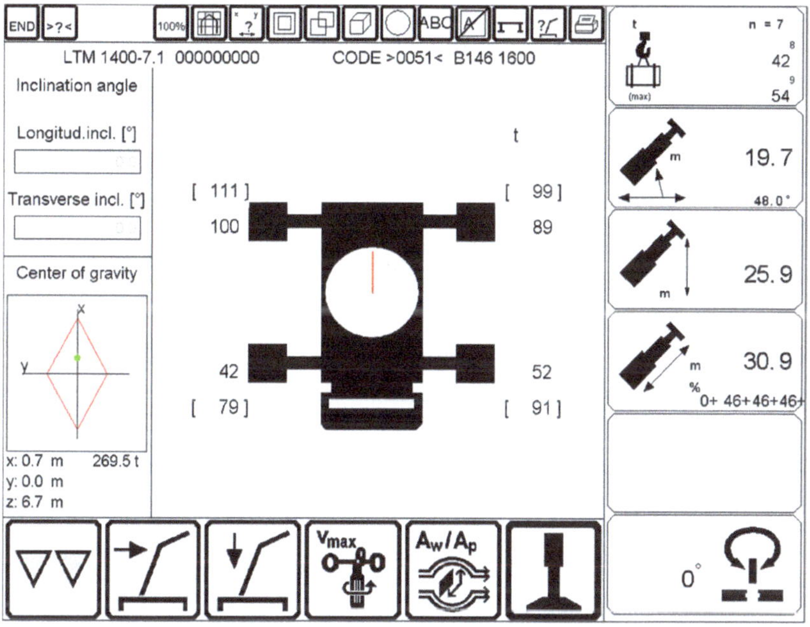

Figure 4-7. Outrigger load calculator printout for a rough terrain crane. Source: Courtesy of Liebherr.

inputting information into the manufacturer's software for loading applied to a bridge structure, it is the engineer's responsibility to verify inputs that reflect the actual equipment and operating conditions.

4.3.4 Miscellaneous Equipment

4.3.4.1 Overview. Contractors usually require additional pieces of miscellaneous equipment to facilitate demolition operations. Examples of common support equipment used in demolition are listed subsequently. Refer to Chapter 3 "Equipment and Tools for Bridge Demolition" for a more detailed description of each piece of equipment:

- Flatbed and/or heavy haul trucks,
- Front end loaders,
- Skid steers,
- Asphalt milling machines, and
- Personnel access equipment.

Design loads for these machines should be based on equipment load data provided by the manufacturer for the specific equipment to be used. Equipment that transports debris should also include any imbalances in the front/rear axle load distribution. Most manufacturers' literature specifies the static tipping load of the machine (a load that will theoretically balance the machine on its front axle). The equipment weight plus any load being picked, applied at the appropriate reach, should be used to statically determine axle loads.

Many of these machines exceed normal AASHTO specifications, as they tend to have short wheelbases and heavy axle loads. When multiple pieces of equipment are used together for demolition, their loading needs to be considered concurrently with all other equipment as a group for combined effects on the structure. It is important to specify in the demolition drawings the size, maximum weight, and any operational or location restrictions associated with these pieces of equipment.

4.3.4.2 Wheel Contact Area. Wheel contact area for rubber-tired equipment may also need to be calculated for analysis. AASHTO specifies a 10 × 20 in. (254 × 508 mm) contact area for typical trucks. The manufacturer's specified contact area should be used whenever possible for demolition equipment. Wheel contact area can be calculated by the wheel load divided by the tire pressure when contact area is not provided.

4.3.5 Dynamic Load Allowance

Dynamic load allowance or impact should be applied to equipment loads to account for the effects of equipment moving or operating on the structure. Minimum impact factor recommendations, based on a survey of demolition engineers across the country, for common removal operations are tabulated subsequently. The values given should be used as minimums in the absence of more specific values provided by the manufacturer, the authority having jurisdiction (DOT or project specification), or as justified by a more detailed analysis.

Demolition Operation	Minimum Impact Factor	Notes
Walking (traveling)	1.05	Demolition equipment is often moving at very slow speeds, and as such, a nominal dynamic allowance is considered appropriate. Wheeled equipment impact should be based on bridge deck conditions and speed constraints
Hoisting and swinging operations	1.05	A larger dynamic factor than listed may be considered to account for potential binding of the member

Concrete removal operations	Hammering—1.25 Shearing—1.10 Panelized slab removal—1.15	Dynamic Allowance during deck removal operations will vary depending on the removal operation. Suggested minimum impact factors given are based on a survey of demolition engineers

4.4 MISCELLANEOUS LOADS

Other types of loading that may be relevant to the structure during demolition operations include environmental loads (snow, ice, wind, and thermal), incidental lateral loads, and locked-in forces. Demolition activities are typically stopped during extreme environmental conditions such as high winds or snow and ice storms. Therefore, it may not be necessary to combine the maximum construction live load with the maximum environmental loads. In addition, demolition equipment may require wind speed operating limits that prevent some construction activities from being performed during extreme events. For areas that are prone to high winds or snow and ice accumulation, consult with the project specifications for any environmental loads to consider.

This is not an exhaustive list of potential loads that may need to be considered. It is ultimately the responsibility of the engineer to determine all relevant loads that should be included in the demolition analysis.

4.4.1 Snow and Ice Loads

Snow and ice loads are typically determined using national or local design specifications. The limited duration of demolition operations, along with staging and control of work, typically allow for the use of reduced load magnitudes and combinations. It is common for the maximum equipment load to control over snow and ice loading. If equipment use is restricted during snow and/or ice events, they are not typically included together as a load combination. However, it is recommended that the demolition plans specify that snow/ice are removed and are not permitted to be stockpiled on the structure.

4.4.2 Wind

Wind loads are typically determined using project specifications or design specifications. FHWA's *Engineering for Structural Stability in Bridge Construction*, ASCE's *Minimum Design Loads and Associated Criteria for Buildings and Other Structures* (ASCE 2017), and AASHTO's *Guide Specifications for Wind Loads on Bridges During Construction* (AASHTO 2017b) are code references commonly used for developing wind loads for temporary demolition conditions. In addition, temporary reduction factors may be used on the basis of the construction period, as specified in ASCE 37 (ASCE 2014). *Engineering for Structural Stability in Bridge Construction* and *Guide Specifications for Wind Loads on Bridges During Construction* also provides guidance for wind loading and shielding effects on girders without concrete decks.

Temporary maximum wind speeds may be specified in the demolition plans for specific stages of the work that can be completed in a short duration, such as a single shift. For these scenarios where the assumed maximum wind speed is used to directly calculate wind pressures, weather conditions can be monitored in advance to plan the working window, and also monitored during the operation to ensure that the work will be completed while meeting the wind speed criteria. Contingency plans are often required in case the work is not completed in the specified timeframe or if weather conditions change abruptly.

During demolition operations, the existing structure may experience stages with a compromised lateral force resisting system. During these temporary stages, the structure may still be subjected to wind loads, and as such, must be checked for both strength and stability.

4.4.3 Incidental or Minimum Lateral Loading

Occasionally, it is appropriate or required by the specifications to apply a minimum lateral load on a structure during demolition operations. AASHTO's *Guide Design Specifications for Bridge Temporary Works outlines Horizontal Construction Loads* (AASHTO 2017a) with the intent that the engineer provides a mechanism for lateral load resistance on an incomplete structure during erection. The same principle applies to the demolition of a structure where the permanent lateral force-resisting system is compromised. These loads could result from a number of factors, including equipment operating (traveling, rotating, braking), personnel, and eccentricities in vertical loading, among others.

These loads are typically calculated on the basis of a percentage of dead or live vertical loads (or a combination of both) and should not be taken as a substitute for environmental loads. See the following table for typical incidental or minimum lateral loading. The engineer may opt to use different values on the basis of project-specific criteria and engineering judgment.

Typical Incidental Lateral Loading Based on Percentages of Vertical Load

Vertical Load	%
Dead	2–10
Live (equipment)	10–20

4.4.4 Locked-In Forces

Locked-in forces or residual forces can occur in a structure through the construction staging used during bridge erection, rehabilitation or retrofitting a bridge, or changes to the supports, such as a pinned connection becoming locked because of corrosion. These forces can vary from project to project and are often determined through advanced analysis. Typically, it is useful to perform upper- and lower-bound analyses to envelope potential effects on the structure. The engineer is responsible for understanding the type, age, and condition of the structure and how these factors may contribute to locked-in or residual forces.

REFERENCES

A1A Software. 2023. 3D lift plan. Accessed December 2, 2023. https://www.3dliftplan.com/Default.aspx

AASHTO (American Association of State Highway and Transportation Officials). 2017a. *Guide design specifications for bridge temporary works*. 2nd ed. Washington, DC: AASHTO.

AASHTO. 2017b. *Guide specifications for wind loads on bridges during construction*. 1st ed. Washington, DC: AASHTO.

AASHTO. 2020. *LRFD bridge design specifications*. 9th ed. Washington, DC: AASHTO.

ASCE. 2014. *Design loads on structures during construction*. ASCE 37-14. Reston, VA: ASCE.

ASCE. 2017. *Minimum design loads and associated criteria for buildings and other structures*. ASCE 7-16. Reston, VA: ASCE.

FHWA (Federal Highway Administration). 2015. *Engineering for structural stability in bridge construction*. Washington, DC: FHWA.

Shapiro, L., and J. Shapiro. 2010. *Cranes and derricks*. 4th ed. New York: McGraw Hill.

Wisconsin Department of Transportation. 2011. *Development of a bridge construction live load analysis guide*. Milwaukee, WI: Wisconsin Department of Transportation.

CHAPTER 5
DEMOLITION WITH EXPLOSIVES

5.1 INTRODUCTION

The use of explosives in bridge demolition has been in vogue for decades. The intent of using explosives for bridge demolition is to bring the structure quickly and safely to the ground where it can be processed with conventional equipment. Explosive demolition on steel structures is completed using shaped charges that cut the steel upon detonation, whereas concrete is typically fractured with explosives to the point that it can no longer support itself.

The goal in any construction operation is to perform the work safely, correctly, and at the lowest cost. Explosive demolition is no different, and bridge demolition is usually engineered to be performed using the least amount of explosive material possible. Fewer charges represent lower material cost, less labor, and less air overpressure (noise); air overpressure can cause damage to surrounding building structures and facilities.

The use of explosives to demolish a bridge is a highly technical process. Significant up-front engineering to determine the charge locations and to analyze the existing structure for preblast activities must take place. The design work is typically divided between two experts, the explosive engineer and the construction engineer (see the following section for more information).

5.2 ROLES FOR THE PROJECT TEAM

5.2.1 General Contractor

General contractors are often hired to oversee and perform many of the larger project construction activities. These include performing feasibility studies to determine whether removal with explosives is right for the project.

Once the decision is made to remove the structure with explosives, the general contractor must work with both the explosives contractor/engineer and the construction engineer to develop the removal plan. This includes determining the size of equipment used for final removal and processing, so as to determine maximum pick weights and sizes. This information is used by the explosives and construction engineers to determine the locations of precuts and shaped charges.

5.2.2 Explosives Contractor/Explosives Engineer

Explosive demolition is typically subcontracted to a specialty explosives contractor. Explosives contractors usually perform the transportation, installation, and detonation of explosives. Explosives contractors may need special federal, state, and even local licenses depending on where the work occurs. Their employees are highly trained and understand the safety protocols that go along with their task. The selection of an experienced, qualified, and competent explosives contractor is paramount to the successful completion of the project.

The explosives contractor will work with an explosives engineer responsible for determining the charge size, type, preferred locations, drop zone studies, and precut or coring requirements to install the charges. The explosives engineer is responsible for determining the timing, staging, and direction of the fallen structure.

5.2.3 Construction Engineer

Once the charge locations are known, a construction engineer with structural analysis experience should be engaged to evaluate the structure for adequacy and stability prior to the blast. This engineer should have experience in this type of work and is not usually involved in the selection of the explosives and direction of fall. The construction engineer is typically hired by the general contractor and must work alongside the explosives contractor to ensure that all critical stages of removal are studied.

Explosive demolition typically requires removal work prior to setting the explosives, such as deck and overlay removal, removal of nonstructural elements, and select removal of secondary structural members. This removal work is done to reduce the amount of explosives required, reduce the amount of processing after demolition, or reduce the weight of the debris. Construction engineers must evaluate the stability of the structure through all removal stages, including modifications such as precutting members or removing parts of the lateral force resisting system, to ensure adequate capacity and structural stability. See the "Structural Engineering Consideration" section for more details.

5.3 MOTIVES FOR SELECTING EXPLOSIVES DEMOLITION

Using explosives to demolish a bridge structure is a highly specialized operation requiring the right combination of site conditions and structure type. It is much less common than conventional demolition, discussed in the previous chapters. Explosive demolition is more comparable to removal of large bridge sections using complex heavy lift operations, such as lowering with strand jacks or removal with SPMTs.

Several reasons to consider the use of explosives for a demolition project includes the following:

1. **Structure Type**: The structure type may present challenges to conventional demolition. Examples could include taller structures where equipment access is difficult, longer spans that are too heavy to be removed with available equipment, or where float-out by barge is not possible. Example structure types include the following: (1) truss bridges, (2) arch bridges (concrete or steel), (3) large area footprint or tall substructures, and (4) long-span plate girder bridges.
2. **Safety**: In some cases, using explosives is the safest method. If performed correctly, spans dropped with explosives will fall straight down. Explosive demolition also has the advantage that personnel can be positioned away from the explosive operation. If the same bridge were to be demolished using conventional methods, personnel would need to be in close proximity to the demolition operation, which could constitute a higher risk.
3. **Condition of the Existing Structure**: Conventional demolition operations often require large equipment to be supported by the existing structure. It is not uncommon for structures to lack the structural capacity to support the necessary equipment loading for conventional demolition.
4. **Cost and Schedule**: Explosive demolition can provide cost and schedule benefits. There are fewer restrictions on the means and methods of removal after the structure has been dropped. Once on the ground, the ability to process the fallen spans is typically limited only by equipment and the manpower to operate them.

The decision to use explosives may be influenced by project specifications. The use of explosives may be prohibited or may require additional discussions and reasons for justification. If explosives are permitted, it is crucial to have early discussions with the owner, engineer, and all stakeholders. A communication plan may be required if the general public will be impacted by closures, restrictions, or noise. People are fearful of explosives; the mere word conjures up an image of flying debris, dust clouds, and broken windows. The use of explosives should be carefully evaluated to ensure that the stakeholders understand the risks and benefits involved. Different segments of the project may lend themselves to the use of explosives, whereas other segments may be too risky and alternative means must be used to demolish them.

5.4 PREPARATION FOR EXPLOSIVES DEMOLITION

5.4.1 Substructure Removal Preparation

5.4.1.1 Pier Implosion/Explosion Overview. Pier removal with explosives is typically justified only in special circumstances. Large concrete piers may be collapsed on themselves (implosion) and steel frame piers can be brought down with external-shaped charges (explosion). For steel-framed piers, preparations would be similar to those described in Section 5.4.2. For concrete piers, preparation measures include drilling the pier so that the demolition subcontractor can set the charges. The demolition subcontractor develops the required explosives and drill locations for each project. Typically, the drilling is minimal in relation to the size of the existing pier and the weakening of the structure does not require further evaluation.

5.4.1.2 Protection of Surrounding Structures. Protection measures are project dependent but may be required for surrounding structures, wildlife, and any potential observers. Flying debris protection can be provided by steel-braided blast mats and/or sandbags set on top of the explosive charges (Figure 5-1). Where available in tidal regions, blasts can be timed to occur during high tide to take advantage of more water cover.

Figure 5-1. Blast mats from protection of surrounding structure during pier implosion.

Bubble curtains have been used for sound mitigation and the protection of wildlife. In special circumstances, cofferdams have been built around the existing piers for both sound and structural protection of surrounding structures. An engineering firm experienced with dynamic finite-element analysis and having a knowledge of blast load effects should be used for designing these special case cofferdam structures.

5.4.2 Superstructure Removal Preparation

5.4.2.1 Drop Zone Considerations. Drop zone is defined as the area directly below the bridge where the debris material is expected to land after the explosives are set off. Drop zones may contain items that are sensitive to falling debris, such as utilities or existing structures. Avoiding these objects often requires specific, surgical removal of portions of the superstructure prior to explosion. Figure 5-2 shows an example of a steel overhang, which was removed prior to explosion to prevent damage to an existing retaining wall directly below. As an alternative, additional systems may be provided to protect sensitive items in the drop zone. These systems

Figure 5-2. Bridge overhangs on top of an existing retaining wall structure. Overhangs had to be removed before dropping the span.

Figure 5-3. Temporary falsework provided above a new pier in a blast zone.

should be designed by a professional engineer as the loading demands will include dynamic impact and may be substantial.

5.4.2.2 Pier Protection from Falling Debris. It is rare for piers to remain in service after a superstructure has been removed with explosives. However, in this case, the explosive charge locations in the superstructure should be located and oriented such that the main structure falls away from the piers. Additional measures should also be taken to protect the pier from smaller flying debris. Piers for adjacent structures may also require protection. If this is the case, it is recommended that temporary protection systems be provided to protect the pier structure and/or portions of the existing structure should be removed prior to explosion to provide additional distance to the dropped structure (Figure 5-3).

5.4.2.3 Removal of Structures for Weight Reduction. Weight reduction prior to dropping a span has many benefits. If the total weight of the dropped structure is reduced, lighter equipment may be used for final removal/processing, which could reduce costs and improve efficiency. If the structure is being dropped into a waterway, this can significantly reduce the pick weight of the debris pieces that have to be retrieved. Reducing the weight of the structure may also result in fewer charge locations, which can lead to a dramatic decrease in cost. Using less explosives also reduces air overpressure (noise) caused by the detonation and minimizes possible damage to nearby windows.

Components commonly removed prior to dropping the span may include asphalt overlay, barriers, utilities, structural deck, and nonessential steel floor system members. Removal of asphalt from the span not only reduces weight, but also eliminates the contamination of asphalt into the debris when the span is further processed.

Structural deck (and nonessential steel framing) removal is common but not necessary. The deck is typically the last thing to be removed prior to dropping the span (Figure 5-4). Removal of deck also significantly weakens and reduces the stiffness of the structure. Decks can be removed using either of the typical demolition methods: rubblized in place or cut into panelized sections and lifted out. Removing the deck in panels is more time-consuming both in field execution and in planning but significantly reduces falling debris and may be required depending on the project location.

5.4.2.4 Precutting Existing Structures. Linear-shaped charges are commonly used for the explosive removal of existing steel structures. Prior to setting the shaped charges, a partial precutting of the structural members is often required to both facilitate installation of the shape charge and limit the amount of explosive charge required to make the final cut with explosives. Limiting explosives can lead to cost savings; it also reduces the air overpressure. Examples of precuts include (1) cutting through the flange sections of built-up box sections so that the charge can be set in the notch directly against the remaining web (Figure 5-5), and (2) cutting through the webs of plate girder sections so that charges can be set against

Figure 5-4. Panelized deck removal by hoisting.

Figure 5-5. Precut double channel box. Flanges cut and shaped charges set against the remaining webs.

the flanges (Figure 5-6). The planning for precuts on complex structures is often an iterative process between the explosive and the construction engineer and contractor as defined in Section 5.2.

Charge locations (and corresponding precut locations) are located to limit the weight of the dropped sections so that removal and processing can be completed with onsite equipment. Larger segments may be dropped when performing explosive demolition over land, as heavy equipment may be used for processing and smaller segments are undesirable (Figure 5-7).

Figure 5-6. Precut web of a plate girder with a shaped charge set against a flange.

Figure 5-7. Truss span blast over land. Blast locations located near the ends of spans. Further processing could occur on ground.
Source: Skanska.

There is a possibility that the dropped structure could become lodged into the existing riverbed when performing explosive demolition over waterways. Based on experience, the initial force required to extract the dropped section may be 125% of the self-weight of the member. This is critical when sizing equipment and determining the shaped charge locations (Figure 5-8).

Figure 5-8. Truss span blast over river and land. More closely spaced blast locations used in the river side to separate into pieces with weights manageable for onsite equipment.

Secondary members, such as lateral wind bracing or cross-frames, are often cut in their entirety prior to dropping the structure with explosives. Full cutting of secondary members helps minimize the explosives charges used and simplifies the blast plan for potential hang-ups. When cutting structural members, it is vital that adequate engineering be performed to ensure that the weakened structure will not unexpectedly collapse prior to demolition. The complete or partial removal of the deck, cuts made to prepare the structure for shaped charges, and cutting of structural elements often result in a significantly weakened lateral load–carrying capacity. Precutting and removal work should be planned and performed to minimize the amount of time the weakened structure is fully exposed to environmental loads.

5.4.2.5 Miscellaneous Nonstructural Preparations. Other preparatory work that should be considered when performing explosives demolition may include the following:

1. Required protection of adjacent bridge or building structures (if applicable). Oftentimes, new bridges are built on a parallel alignment only a few yards away from the existing span. When shaped charges detonate, small pieces of shrapnel may be ejected. The adjacent bridge may need to be protected from flying debris, depending on its proximity and construction.
2. Preparation and consideration of surrounding buildings and structures (windows, etc.). Surrounding structures may be outside of the immediate blast zone where flying debris is not a concern, but still within the range of air overpressure effects. In some cases, this could be as simple as boarding up windows prior to the demolition. Reduction of shape charge size can also help in reducing the amount of damage to surrounding buildings.
3. Preconstruction survey. A preconstruction survey is highly recommended to thoroughly document the condition of nearby properties, including documenting previously broken windows, cracks in foundations, damage to existing building facades, and others. This type of survey is commonly required in many contracts and is especially important for explosives demolition. This helps in monitoring and repairing actual damage and helps prevent fraudulent claims of damage caused by this type of demolition.
4. Environmental considerations. Environmental considerations vary by location but may include noise protection, contamination of surrounding areas, or impacts on wildlife habitats.
5. Public outreach. Educating project stakeholders can help reduce concerns associated with explosives. Informing the public, project owners, and resident engineers on the variety of means and methods

that could be used to demolish a bridge can result in the realization that explosives demolition is the least intrusive, safest means to demolish the span.

6. Evaluation of underground utilities and infrastructure. Structures such as tunnels and roadways adjacent to explosive demolition sites should also be inspected and their condition documented prior to explosives demolition. The impact of the dropped structure could result in damage to underground utilities and should be evaluated prior to the operation.

5.5 STRUCTURAL ENGINEERING CONSIDERATIONS

5.5.1 Deck/Floor System Removal

The existing structure should be rated by an engineer for any equipment required to remove the deck or other members. Key considerations include (1) reduction in structural capacity, (2) stress reversals in members, connections, and bearings, and (3) a condition assessment of the existing structure. Removal of the deck should be done before the precuts are made for the shaped charges. See Chapter 4 "Loads" for a discussion of equipment load development.

5.5.2 Main Load–Carrying Members with Precuts

Main load–carrying members are often precut to allow for the proper installation of shaped charges. Evaluation of the reduced capacities of the precut members should be performed by a qualified structural engineer. Any equipment live load allowed on the existing structure after precutting should be carefully evaluated. The study of member adequacy should consider localized stresses and local buckling between precuts, as well as global member demands. These evaluations can be made using typical code provisions or can be analyzed using a finite-element model (Figure 5-9). The size and location of the precuts should be clearly indicated on the demolition plans, along with the staging and any restrictions or limitation on the bridge during or after the precutting operations.

5.5.3 Secondary Bracing Members with Precuts

Wind bracing or lateral bracing is often precut entirely prior to dropping spans with explosives. This typically requires the evaluation of alternative load paths needed to resist lateral loads. It can also result in the requirement that the entire operation be performed during a controlled wind period with low forecasted wind speeds. Contingency plans, which may include

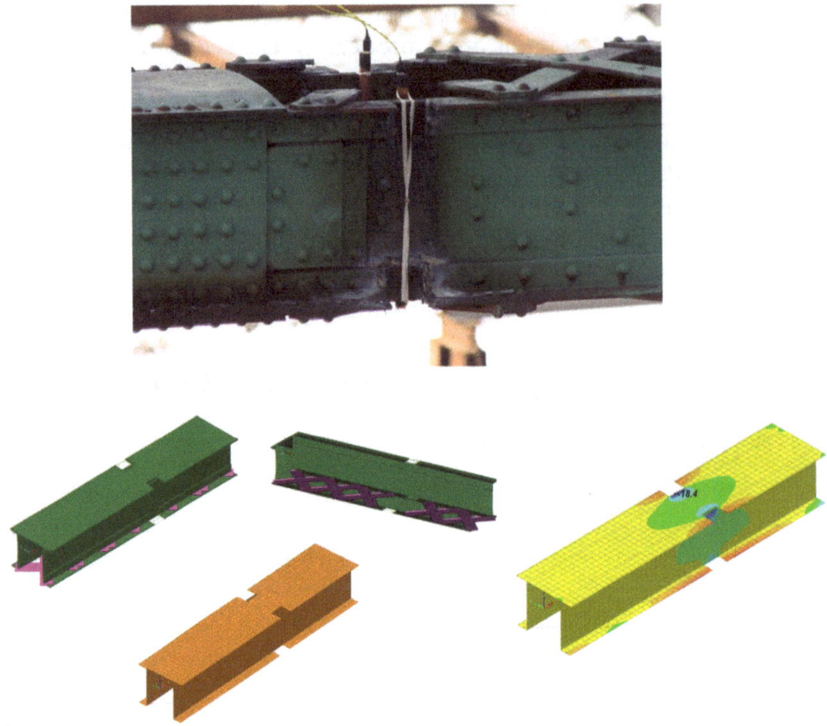

Figure 5-9. Precut built-up box section with finite-element analysis.

temporary bracing, should be developed to account for any delays in the operation or unexpected wind events. This is of particular importance for longer spans [> 300 ft (91.5 m)]. The assumptions and limitations on the structure should be stated in the demolitions plans, including the assumed exposure duration, the maximum wind speed (if applicable), and contingency plans (if required).

5.5.4 Load Effects on the Remaining Structure

If a superstructure is removed in sections and portions of the structure are to remain after using explosives (i.e., for continuous span structures removed in phases), load effects from instantaneous loss of load must be considered on the remaining structure. One method is to apply a force on the remaining structure equivalent to two times the static force in the cut members prior to removal. This force should be applied in the opposite direction as the internal force to account for the dynamic rebound of the sudden loss of force. This method is consistent with evaluating dynamic cable loss in *Recommendations for Stay Cable Design, Testing, and Installation*

(PTI 2018), which describes a simplified static load caused by the sudden loss of a cable in a structure. Alternatively, a dynamic, nonlinear finite-element model may be designed for more accurate dynamic effects from the instantaneous loss of load.

The 2× load assumption can produce conservative results, often with costly hold-downs to maintain stability of the remaining structure. Nonlinear dynamic models will produce more realistic results, often indicating load reversal, and will reveal absolute maximum and minimum forces for design. Blast effects on the surrounding structures should be evaluated by an explosive engineer.

REFERENCE

PTI (Post-Tensioning Institute). 2018. *Recommendations for stay cable design, testing, and installation*. PTI Committee DC-45: Cable-Stayed Bridge. Farmington Hills, MI: PTI.

APPENDIX
BRIDGE DEMOLITION PLANS SUGGESTED CHECKLIST

Item	Included	N/A
Name and credentials of the demolition supervisor	☐	☐
Assumed loading conditions	☐	☐
Maximum wind speed including any assumptions about duration-based reduction factors	☐	☐
Construction live load	☐	☐
Equipment loading	☐	☐
Other environmental loads (ice, seismic, stream flow, etc.)	☐	☐
Impact factors	☐	☐
Site plan showing	☐	☐
Adjacent structures	☐	☐
Underground utilities	☐	☐
Overhead utilities	☐	☐
Roads, railroads, waterways, and clearances	☐	☐
Framing plan showing orientation and member labels (special highlight for fixed/free bearings)	☐	☐

Demolition sequence	☐	☐
Crane location, working radius, and pick weights	☐	☐
Crane travel paths, work pads, foundations	☐	☐
Location and sequence for temporary bracing and blocking	☐	☐
Location and sequence for temporary support	☐	☐
Drawings for all temporary bracing, blocking, and/or temporary supports	☐	☐
Design loads	☐	☐
Required capacity or safe working load	☐	☐
Elevations and details	☐	☐
Foundation requirements	☐	☐
Installation/removal procedures	☐	☐
Lift plans (critical lift plans where required)	☐	☐
Crane configuration including radius, counterweight, boom length	☐	☐
Pick weight including rigging weight	☐	☐
Center of gravity of a lifted piece	☐	☐
Rigging system arrangement	☐	☐
Rigging system sizes or minimum rated capacity	☐	☐
Drawings for all jacking operations	☐	☐
Location of jacks	☐	☐
Strengthening details	☐	☐
Providing a jacking table with loads, jack pressure, movement/elongation	☐	☐
Drawings for post-tensioning procedures and requirements	☐	☐
Deck removal methods	☐	☐
Traffic control plan	☐	☐
Contingency plans	☐	☐
Location of debris storage and disposal	☐	☐

Protection measures for existing structures (railroad tracks, roadways) and utilities	☐	☐
Protection of waterways or environmentally sensitive areas	☐	☐
Predemolition and postdemolition surveys	☐	☐
Noise pollution/mitigation	☐	☐
Dust control	☐	☐
Water/runoff control	☐	☐
Evaluating existing conditions on the basis of an inspection or rating report	☐	☐
Assumed material strength for rolled shapes, bolts/rivets, concrete, rebar	☐	☐

INDEX

Papier | Fördert gute Waldnutzung
FSC® C083411

Zeitfracht Medien GmbH
Ferdinand-Jühlke-Straße 7
99095 Erfurt, Deutschland
produktsicherheit@kolibri360.de